Nanostructured Materials based on Noble Metals for Advanced Biological Applications

Nanostructured Materials based on Noble Metals for Advanced Biological Applications

Special Issue Editor

Iole Venditti

MDPI • Basel • Beijing • Wuhan • Barcelona • Belgrade

Special Issue Editor
Iole Venditti
Roma Tre University
Italy

Editorial Office
MDPI
St. Alban-Anlage 66
4052 Basel, Switzerland

This is a reprint of articles from the Special Issue published online in the open access journal *Nanomaterials* (ISSN 2079-4991) from 2018 to 2019 (available at: https://www.mdpi.com/journal/nanomaterials/special_issues/noble_metals_nano).

For citation purposes, cite each article independently as indicated on the article page online and as indicated below:

LastName, A.A.; LastName, B.B.; LastName, C.C. Article Title. *Journal Name* **Year**, *Article Number*, Page Range.

ISBN 978-3-03928-833-5 (Pbk)
ISBN 978-3-03928-834-2 (PDF)

Contents

About the Special Issue Editor . . . vii

Preface to "Nanostructured Materials based on Noble Metals for Advanced Biological Applications" . . . ix

Iole Venditti
Nanostructured Materials Based on Noble Metals for Advanced Biological Applications
Reprinted from: *Nanomaterials* **2019**, *9*, 1593, doi:10.3390/nano9111593 . . . 1

Carlota Auría-Soro, Tabata Nesma, Pablo Juanes-Velasco, Alicia Landeira-Viñuela, Helena Fidalgo-Gomez, Vanessa Acebes-Fernandez, Rafael Gongora, María Jesus Almendral Parra, Raúl Manzano-Roman and Manuel Fuentes
Interactions of Nanoparticles and Biosystems: Microenvironment of Nanoparticles and Biomolecules in Nanomedicine
Reprinted from: *Nanomaterials* **2019**, *9*, 1365, doi:10.3390/nano9101365 . . . 6

Ilaria Fratoddi, Iole Venditti, Chiara Battocchio, Laura Carlini, Simone Amatori, Marina Porchia, Francesco Tisato, Federica Bondino, Elena Magnano, Maura Pellei and Carlo Santini
Highly Hydrophilic Gold Nanoparticles as Carrier for Anticancer Copper(I) Complexes: Loading and Release Studies for Biomedical Applications
Reprinted from: *Nanomaterials* **2019**, *9*, 772, doi:10.3390/nano9050772 . . . 26

Youngjin Jang, Nohyun Lee, Jeong Hyun Kim, Yong Il Park and Yuanzhe Piao
Shape-Controlled Synthesis of Au Nanostructures Using EDTA Tetrasodium Salt and Their Photothermal Therapy Applications
Reprinted from: *Nanomaterials* **2018**, *8*, 252, doi:10.3390/nano8040252 . . . 39

Jinhyung Lee, Eun-Ah You, Do Won Hwang, Shinill Kang and Jung-Sub Wi
Active Accumulation of Spherical Analytes on Plasmonic Hot Spots of Double-Bent Au Strip Arrays by Multiple Dip-Coating
Reprinted from: *Nanomaterials* **2019**, *9*, 660, doi:10.3390/nano9050660 . . . 50

Alexandra-Cristina Burdușel, Oana Gherasim, Alexandru Mihai Grumezescu, Laurențiu Mogoantă, Anton Ficai and Ecaterina Andronescu
Biomedical Applications of Silver Nanoparticles: An Up-to-Date Overview
Reprinted from: *Nanomaterials* **2018**, *8*, 681, doi:10.3390/nano8090681 . . . 58

Paolo Prosposito, Luca Burratti, Arianna Bellingeri, Giuseppe Protano, Claudia Faleri, Ilaria Corsi, Chiara Battocchio, Giovanna Iucci, Luca Tortora, Valeria Secchi, Stefano Franchi and Iole Venditti
Bifunctionalized Silver Nanoparticles as Hg^{2+} Plasmonic Sensor in Water: Synthesis, Characterizations, and Ecosafety
Reprinted from: *Nanomaterials* **2019**, *9*, 1353, doi:10.3390/nano9101353 . . . 83

Federica Rinaldi, Elena del Favero, Johannes Moeller, Patrizia Nadia Hanieh, Daniele Passeri, Marco Rossi, Livia Angeloni, Iole Venditti, Carlotta Marianecci, Maria Carafa and Ilaria Fratoddi
Hydrophilic Silver Nanoparticles Loaded into Niosomes: Physical–Chemical Characterization in View of Biological Applications
Reprinted from: *Nanomaterials* **2019**, *9*, 1177, doi:10.3390/nano9081177 . . . 99

About the Special Issue Editor

Iole Venditti serves as Inorganic Chemist at the Department of Sciences in Roma Tre University of Rome, Italy. She received her Ph.D. in Materials Science from Sapienza University of Rome. Her research focuses on micro- and nanostructured materials based on noble metals (gold, platinum and silver) or polymers. These materials have found application in optoelectronic devices, drug delivery, and environmental pollutants sensing. Iole has cooperated with many researchers worldwide on various research projects on these topics. She serves as Editor for several top-ranked international journals, including *Nanomaterials*, *Polymers*, and *Chemosensors*. To date, she has 2 patents, has collaborated in 2 books, and has published 83 scientific papers, with more than 1800 citations and an h-index of 28, according to Scopus.

Preface to "Nanostructured Materials based on Noble Metals for Advanced Biological Applications"

This book is my positive answer to a call I received from *Nanomaterials*. In my opinion the topic "Nanostructured Materials Based on Noble Metals for Advanced Biological Applications" is a hotspot of the international and interdisciplinary research between several emerging and advanced applications fields, such as nanosensors and nanomedicine. It is nowadays generally accepted that nanostructured noble metals allow the production of highly competitive materials. In fact, a specific design and rather simple and reliable preparation techniques can be used to obtain optimized material uses and possibilities for their reusability. One expects amazing future developments for these nanotechnologies from research laboratories to key industrial areas. I consequently accepted, with pleasure, the invitation to serve as Guest Editor for this Special Issue, that presents some advanced methods of gold and silver nanomaterials preparation/synthesis as well as their innovative applications. The book opens with an editorial that presents the various works and their common thread. The first contribution is a review broadly based on metallic nanoparticles in the biological field followed by more specific works on gold nanoparticles to be used in various advanced applications, such as theragnostics. Following is a study on the current state-of-the-art on the use of silver nanoparticles in nanomedicine. At the end of this collection of research, we find silver nanoparticles used to develop applications in the environmental and biotechnology fields. The resulting volume proved to be international, with authors from several countries (France, Germany, Korea, Israel, Italy, Romania, Spain). All contributions represent works from prominent universities and research centers all over the world. I, as Guest Editor, and the MDPI staff are pleased to offer this Special Issue to all interested readers to the wide world of nanomaterials.

Iole Venditti
Special Issue Editor

nanomaterials

Editorial

Nanostructured Materials Based on Noble Metals for Advanced Biological Applications

Iole Venditti

Sciences Department, Roma Tre University, via della Vasca Navale 79, 00146 Rome, Italy; iole.venditti@uniroma3.it; Tel.: +39-06-5733-3388

Received: 23 October 2019; Accepted: 7 November 2019; Published: 10 November 2019

Abstract: This special issue focuses on highlighting the progress of last decade regarding the new nanostructured materials based on noble metals, especially gold and silver. Innovative preparations, functionalizations, and characterizations of these nanomaterials are investigated. Moreover, biotechnological applications, and advanced uses of these compounds for environmental sensing are reported. In particular gold and silver nanomaterials are widely studied due to their high stability, amazing chemical–physical features and, for silver, marked antibacterial properties. It is also hoped that the current special issue will encourage multidisciplinary research on noble metal nanomaterials, expanding the range of potential biological applications. This must be associated with improvements in synthetic methods and with economic feasibility studies of the proposed processes, also exploring the ecotoxicological aspects.

Keywords: nanomaterials; noble metal nanoparticles; gold nanomaterials; silver nanomaterials; hybrid metal–polymer nanoparticles; nanomedicine; biotechnological applications; nanomaterials for drug delivery; nanomaterials for sensing

---

Nanostructured materials based on noble metal nanoparticles (NPs) allow synergistic enhancement of their functional properties due to the low dimensionality. In fact, nanodimension is the strategic key for a wide range of bio-applications, such as biosensors, biocatalysis, drug delivery, imaging, and theranostic applications [1–5]. A huge variety of new materials and composites have been improved, mainly via chemical approaches, using metal surface engineering to build new synergic hybrid systems, in particular based on gold and silver nanoparticles (AuNPs, AgNPs) [6–8].

This special issue focuses on highlighting the progress of new nanostructured materials, based on noble metals, their preparation, functionalization, characterization, and advanced application in biological fields. In fact, in this last decade, gold and silver nanoparticles have been widely used in advanced biological technologies, thanks to the high stability of the former and to the marked antibacterial properties of the latter.

Burdușel et al. [9], make a review about nanostructured silver compounds intensively explored for unconventional and enhanced biomedical applications, thanks to their size-related attractive physicochemical properties and biological functionality, including their high antimicrobial efficiency and non-toxic nature. AgNP-based nanosystems and nanomaterials are suitable alternatives for drug delivery, wound dressing, tissue scaffold, and protective coating applications. Various physicochemical parameters were related to the intrinsic antimicrobial effects exhibited by AgNPs, such as size, shape, concentration, surface charge, and colloidal state. Moreover, the impressive available surface of nanosilver allows the coordination of many ligands, thus enabling tremendous possibilities with respect to the surface functionalization of AgNPs.

There is a significant amount of research data proving the beneficial effects of AgNPs in novel biocompatible and nanostructured materials and devices developed for modern therapeutic strategies. In addition to their attractive and versatile antimicrobial potential, AgNPs provide additional mechanical, optical, chemical, and biological peculiarities that recommend them for the design, obtaining, evaluation, and clinical assessment of performance-enhanced biomaterials and medical devices. Still, thorough investigations regarding their short-term and long-term toxicity, as well as the responsible toxic-related mechanisms, are required.

The current limitations related to conventional healthcare practice and the latest challenges resulting from nanosilver-based technology outline the impressive potential of silver nanoparticles in biomedicinal applications. Whether we consider the modification of available biomaterials and devices or the development of novel nanostructured ones, AgNPs are ideal candidates for achieving the very close modern biomedicine desideratum.

Auría-Soro et al. present a review on the significant impact of nanotechnology on medicine [10]. The studies about the biological response to NPs are greatly investigated in parallel with nano-bio interactions, which have influenced NP design. Both were in concordance with the evolution of NPs for biomedical applications. Many studies have investigated and demonstrated that NPs can enter into the human system. Therefore, the NP characteristics on biological systems, such as their physicochemical properties (size, shape, surface, coating and morphology, surface charge, hydrophobicity, chemical composition, structure, and the state of agglomeration), the types of biomolecules present, and the bio-identity of NP protein corona are important issues to characterize in order to know how they interact with cells, organisms, biological medium, biomolecules, and other biological systems or even with other nanomaterials. These studies helped determine their possible biocompatibility and toxicity in biological micro-environments and to engineer nontoxic nanomaterials, which may be used in biomedical applications.

With the potentially wide application of NPs in the future, these may be extensively used in various fields, especially in immunotherapy for clinical diagnosis and therapy based on their size, biocompatibility, surface chemistry, and adjustable toxicity. Immunotherapy combined with nanomedicines may be used to treat different types of cancer due to their excellent efficacy in penetration, specific retention, and killing of tumor cells.

The human proteome study [11] can be an arduous and discouraging task due to the high number of proteins, encoded by around 25,000 different genes, from which multiple protein variants are generated by post-translational modifications. The concept of proteomics involves a comprehensive study on the structures, localizations, post-translational modifications, functions, and interactions of all proteins expressed by an organism at a certain time and under certain conditions. The nanotechnology field has been expanded by providing innovative methods capable of responding to proteomic demands. In this sense, nanotechnology applications in proteomics have established a novel technical platform termed "nanoproteomics." Detection techniques without labels are useful in the study of protein interaction kinetics, thanks to avoiding steric impediments caused by the presence of labels. The design and development of new multi-functional platforms based on nanomedicine could be of great interest in the unlabeled detection of protein–protein interactions given the possibility of synthesizing de novo proteins "in vitro" in the presence of these nanosystems.

Jang et al. investigate a facile and effective shape-controlled synthesis of gold nanostructures and their photothermal therapeutic effect [12]. The described procedure involves the simple mixing of tetrachloroauric acid ($HAuCl_4$) and Ethylenediaminetetraacetic acid tetrasodium salt (EDTA tetrasodium salt) in an aqueous solution at room temperature, without additional ligands or toxic reagents. Adjusting the molar ratios of $HAuCl_4$ to EDTA tetrasodium salt enables effective morphology control of Au nanostructures from spheres to branched forms and nanowire networks. Detailed control experiments revealed that the four deprotonated carboxylic acids of the EDTA tetrasodium salt provided effective growth control and stabilization. The Au nanowire networks showed strong absorption in the near-infrared (NIR) region and hence were suitable for photothermal therapy. Under NIR irradiation, the Au nanowire networks allowed for selective destruction of cancerous human primary glioblastoma cells (U87MG cells) by local heating, generated by the NIR absorption. This work demonstrates the development of a simple synthetic route to NIR-active Au nanostructures, which can be extended to other applications including optical sensing and surface-enhanced Raman scattering.

For effective placement of target analytes on sensor surfaces and their monitoring, Lee et al. present a straightforward measurement method based on nanoplasmonic sensors of double-bent Au strip (DAS) arrays, multiple dip-coating of analytes, self-alignment of analytes in the region of strong plasmonic fields, and spectrometric monitoring of localized surface plasmon resonance (LSPR) peaks [13]. Using this method, closely packed polystyrene (PS) beads in the valleys of the DAS array and pH-dependent stability of the exosomes were successfully monitored in terms of shifts in the LSPR peaks. As a small amount of target analytes can accumulate in the plasmonic hot spot due to multiple dip-coating cycles, and the LSPR peak can be measured with a conventional UV-vis (ultraviolet–visible) spectrometer under physiological conditions, it is expected that the proposed measurement platform will be useful for studying the stability of various drug delivery vesicles and their efficiencies.

Fratoddi et al. prepare conjugates between strongly hydrophilic gold nanoparticles, AuNPs, and copper(I) complexes [14]. In particular, loading and release studies were performed using two different copper(I) antitumor complexes, namely $[Cu(PTA)_4]^+[BF_4]^-$ (A; PTA = 1,3,5-triaza-7-phosphadamantane) and $[HB(pz)_3Cu(PCN)]$ (B; $HB(pz)_3$ = tris(pyrazolyl)borate, PCN = tris(cyanoethyl)phosphane). In the water-soluble compound A, the metal is tetrahedrally arranged in a cationic moiety, while compound B is a mixed-ligand (scorpionate/phosphane), neutral complex insoluble in water. Loading protocols and efficiency are also related to these structural aspects and were optimized to obtain loading efficiency $\eta = 90 \pm 4\%$ and $\eta = 65 \pm 10\%$, respectively, for AuNPs-A and AuNPs-B. Structural differences of A and B induced different behaviors regarding the interactions with the gold surface, as showed by the high-resolution X-ray photoelectron spectroscopy (HR-XPS) studies. In fact, for compound A, nitrogen partially transfers electrons to the surface of the metal nanoparticles, creating an interaction that causes a slow release in water, less than 10% in 4 days. On the other hand, in B compound the N≡C−R groups hook onto the surface of the gold, producing a strong interaction that makes the release not appreciable in the same time interval (up to 4 days). Therefore, both AuNPs-A and AuNPs-B represent promising examples of water-soluble gold nanocarriers suitable to improve the bioavailability of synthetic drugs, especially considering the enhanced permeability and retention (EPR) effect of AuNPs. AuNPs-A, which achieved a slow release, opens the way for biological in vitro studies to explore the synergic activity of copper complexes and gold nanoparticles.

Rinaldi et al. load hydrophilic AgNPs in two different niosomes, producing two systems, namely NioTw20 + AgNPs and NioSp20 + AgNPs [15]. A deep physical–chemical characterization was carried out to obtain information on the influence of AgNPs on the preparation and features of niosomal formulations. First of all, the dynamic light scattering (DLS) studies confirm the nanosize and stability of both systems in water. Moreover, the entrapment efficiency for the two systems was investigated, and it was more efficient for Span 20 than Tw20 niosomes, which was probably related to their different internal structures. Microviscosity and polarity investigations demonstrated that no interactions occurred between the niosomal double layer and the AgNPs, which were probably located inside aqueous compartments. The small-angle X-ray scattering (SAXS) data confirmed the presence of the AgNPs located inside the aqueous compartment of the two niosomal systems, and also allowed highlighting the different structures of their double layers. The morphological characterization indicates that the niosomes maintained spherical shapes. Moreover, stability was confirmed in water, bovine serum, and human serum. Moreover, hydrophilic and lipophilic probe release profiles were obtained in 4-(2-hydroxyethyl)-1-piperazineethanesulfonic acid (HEPES) and in human serum. In conclusion, both systems evidenced the entrapment of AgNPs: NioTw20-AgNPs and NioSp20-AgNPs. The two systems are stable in water, bovine serum, and human serum, and maintain the ability to entrap also hydrophilic or lipophilic model molecules. This work demonstrates that the niosomes' features are not altered by AgNPs loading and confirms that these niosomal formulations are good candidates for the delivery of AgNPs together with other drugs, opening new promising ways for their biotechnological applications.

In the last article of the special issue, silver nanoparticles are investigated as sensing materials for Hg^{2+} in water [16]. AgNPs were synthesized using hydrophilic capping agents, i.e., citrate (Cit) and L-cysteine (L-cys). The characterization by means of UV-vis, Fourier transform infrared spectroscopy (FTIR), HR-XPS and near edge X-ray absorption fine structure (NEXAFS) spectroscopies, confirmed the surface functionalization. Their nanodimensions were studied by DLS and transmission electron microscopy (TEM) analysis, showing diameter less than 10 nm. AgNPs showed high selectivity and sensitivity for Hg^{2+} in water (concentration range: 1–10 ppm) respect to 16 different metal ions investigated. The AgNPs-Hg^{2+} system was deeply investigated by means of DLS, inductively coupled plasma mass spectrometry (ICP-MS), TEM and HR-XPS. Measurements of Ag concentration in fresh and marine aqueous media showed low Ag^{+} ions release, probably due to the good Cit/L-cys covering, also confirmed by HR-XPS data. Moreover, AgNPs ecosafety was confirmed by ecotoxicity tests which showed no effects on algal growth of both freshwater *R. subcapitata* and marine *P. tricornutum* algae in the range of tested concentrations (10–500 mg/L). Our results further support the hypothesis that this specific coating of AgNPs prevent dissolution of Ag^{+} ions in both fresh and saltwater. These results open new ways for AgNPs sensing applications in environmental tests on more complex biological systems, up to tests on real environmental aquatic scenarios.

In conclusion, as editor of this special issue, I am aware that the diversity and innovation of new compounds and tools that are rapidly developing in the field of multidisciplinary research related to nanomaterials based on noble metals, cannot all be collected in a single volume. However, I am sure that this collection will contribute to the interest of research in this area, providing our readers with a broad and updated scenario on this topic.

Funding: This research received no external funding.

Acknowledgments: A special thank you to all the authors for submitting their studies to the present Special Issue and for its successful completion. I deeply acknowledge the Nanomaterials reviewers and editorial staff for enhancing the quality and impact of all submitted papers. Moreover, the Grant of Excellence Departments, MIUR-Italy (ARTICOLO 1, COMMI 314337 LEGGE 232/2016) is gratefully acknowledged.

Conflicts of Interest: The author declares no conflict of interest.

References

1. Maccora, D.; Dini, V.; Battocchio, C.; Fratoddi, I.; Cartoni, A.; Rotili, D.; Castagnola, M.; Faccini, R.; Bruno, I.; Scotognella, T.; et al. Gold nanoparticles and nanorods in nuclear medicine: A mini review. *Appl. Sci.* **2019**, *9*, 3232. [CrossRef]
2. Venditti, I. Gold nanoparticles in photonic crystals applications: A review. *Materials* **2017**, *10*, 97. [CrossRef] [PubMed]
3. Venditti, I. Engineered gold-based nanomaterials: Morphologies and functionalities in biomedical applications. A mini review. *Bioengineering* **2019**, *6*, 53. [CrossRef] [PubMed]
4. Prosposito, P.; Mochi, F.; Ciotta, E.; Casalboni, M.; Venditti, I.; Fontana, L.; Testa, G.; Fratoddi, I. Hydrophilic silver nanoparticles with tunable optical properties: Application for the detection of heavy metals in water. *Beilstein J. Nanotechnol.* **2016**, *7*, 1654. [CrossRef] [PubMed]
5. Loiseau, A.; Asila, V.; Boitel-Aullen, G.; Lam, M.; Salmain, M.; Boujday, S. Silver-based plasmonic nanoparticles for and their use in biosensing. *Biosensors* **2019**, *9*, 78. [CrossRef] [PubMed]
6. Darabdhara, G.; Das, M.R.; Singh, S.P.; Rengan, A.K.; Szunerits, S.; Boukherroub, R. Ag and Au nanoparticles/reduced graphene oxide composite materials: Synthesis and application in diagnostics and therapeutics. *Adv. Colloid Interface Sci.* **2019**, *271*, 101991. [CrossRef] [PubMed]
7. Tan, H.-L.; Teow, S.-Y.; Pushpamalar, J. Application of metal nanoparticle–hydrogel composites in tissue regeneration. *Bioengineering* **2019**, *6*, 17. [CrossRef] [PubMed]
8. Silva, C.O.; Pinho, J.O.; Lopes, J.M.; Almeida, A.J.; Gaspar, M.M.; Reis, C. Current trends in cancer nanotheranostics: Metallic, polymeric, and lipid-based systems. *Pharmaceutics* **2019**, *11*, 22. [CrossRef] [PubMed]

9. Burdușel, A.-C.; Gherasim, O.; Grumezescu, A.M.; Mogoantă, L.; Ficai, A.; Andronescu, E. Biomedical Applications of Silver Nanoparticles: An Up-to-Date Overview. *Nanomaterials* **2018**, *8*, 681. [CrossRef] [PubMed]
10. Auría-Soro, C.; Nesma, T.; Juanes-Velasco, P.; Landeira-Viñuela, A.; Fidalgo-Gomez, H.; Acebes-Fernandez, V.; Gongora, R.; Parra, M.J.A.; Manzano-Roman, R.; Fuentes, M. Interactions of Nanoparticles and Biosystems: Microenvironment of Nanoparticles and Biomolecules in Nanomedicine. *Nanomaterials* **2019**, *9*, 1365. [CrossRef] [PubMed]
11. Jia, L.; Lu, Y.; Shao, J.; Liang, X.; Xu, Y. Nanoproteomics: A New Sprout from Emerging Links between Nanotechnology and Proteomics. *Trends Biotechnol.* **2013**, *31*, 99. [CrossRef] [PubMed]
12. Jang, J.; Lee, N.; Kim, J.H.; Park, Y.I.; Piao, Y. Shape-Controlled Synthesis of au Nanostructures Using EDTA Tetrasodium Salt and Their Photothermal Therapy Applications. *Nanomaterials* **2018**, *8*, 252. [CrossRef] [PubMed]
13. Lee, J.; You, E.-A.; Hwang, D.W.; Kang, S.; Wi, J.-S. Active Accumulation of Spherical Analytes on Plasmonic Hot Spots of Double-Bent Au Strip Arrays by Multiple Dip-Coating. *Nanomaterials* **2019**, *9*, 660. [CrossRef] [PubMed]
14. Fratoddi, I.; Venditti, I.; Battocchio, C.; Carlini, L.; Porchia, M.; Tisato, F.; Bondino, F.; Magnano, E.; Pellei, M.; Santini, C. Highly hydrophilic gold nanoparticles as carrier for anticancer copper (I) complexes: Loading and release studies for biomedical applications. *Nanomaterials* **2019**, *9*, 772. [CrossRef] [PubMed]
15. Rinaldi, F.; del Favero, E.; Moeller, J.; Hanieh, N.; Passeri, D.; Rossi, M.; Angeloni, L.; Venditti, I.; Marianecci, C.; Carafa, I.; et al. Hydrophilic silver nanoparticles loaded into niosomes: Physical-chemical characterization in view of biological applications. *Nanomaterials* **2019**, *9*, 1177. [CrossRef] [PubMed]
16. Prosposito, P.; Buratti, L.; Bellingeri, A.; Protano, G.; Falerid, C.; Corsi, I.; Battocchio, C.; Iucci, G.; Tortora, L.; Secchi, V.; et al. Biofunctionalized silver nanoparticles as Hg^{2+} plasmonic sensor in water: Synthesis characterizations and ecosafety. *Nanomaterials* **2019**, *9*, 1353. [CrossRef] [PubMed]

nanomaterials

MDPI

Review

Interactions of Nanoparticles and Biosystems: Microenvironment of Nanoparticles and Biomolecules in Nanomedicine

Carlota Auría-Soro [1,2], Tabata Nesma [1], Pablo Juanes-Velasco [1], Alicia Landeira-Viñuela [1], Helena Fidalgo-Gomez [1], Vanessa Acebes-Fernandez [1], Rafael Gongora [1], María Jesus Almendral Parra [2], Raúl Manzano-Roman [3,*] and Manuel Fuentes [1,3,*]

1 Department of Medicine and General Cytometry Service-Nucleus, Cancer Research Centre (IBMCC/CSIC/USAL/IBSAL), 37007 Salamanca, Spain; idu007523@usal.es (C.A.-S.); tabata.nesma@gmail.com (T.N.); pablojuanesvelasco@usal.es (P.J.-V.); alavi29@usal.es (A.L.-V.); hfidalgogomez@gmail.com (H.F.-G.); vanessaacebes@usal.es (V.A.-F.); rgongora@usal.es (R.G.)

2 Department of Analytical Chemistry, Nutrition and Food Science, Faculty of Chemistry, University of Salamanca, 37008 Salamanca, Spain; almendral@usal.es

3 Proteomics Unit. Cancer Research Centre (IBMCC/CSIC/USAL/IBSAL), 37007 Salamanca, Spain

* Correspondence: rmanzano@usal.es (R.M.-R.); mfuentes@usal.es (M.F.); Tel.: +34-923294811 (M.F.)

Received: 15 July 2019; Accepted: 18 September 2019; Published: 24 September 2019

Abstract: Nanotechnology is a multidisciplinary science covering matters involving the nanoscale level that is being developed for a great variety of applications. Nanomedicine is one of these attractive and challenging uses focused on the employment of nanomaterials in medical applications such as drug delivery. However, handling these nanometric systems require defining specific parameters to establish the possible advantages and disadvantages in specific applications. This review presents the fundamental factors of nanoparticles and its microenvironment that must be considered to make an appropriate design for medical applications, mainly: (i) Interactions between nanoparticles and their biological environment, (ii) the interaction mechanisms, (iii) and the physicochemical properties of nanoparticles. On the other hand, the repercussions of the control, alter and modify these parameters in the biomedical applications. Additionally, we briefly report the implications of nanoparticles in nanomedicine and precision medicine, and provide perspectives in immunotherapy, which is opening novel applications as immune-oncology.

Keywords: nanoparticles; interactions; protein corona; nanomedicine; biomolecules

1. Introduction

One definition of nanotechnology comes from the statement by the US National Science and Technology Council [1], which states: "The essence of nanotechnology is the ability to work at the molecular level, atom by atom, to create large structures with fundamentally new molecular organization. The aim is to exploit these properties by gaining control of structures and devices at atomic, molecular, and supramolecular levels and to learn to efficiently manufacture and use these devices." Other authors describe nanotechnology as the combinatorial study and integration of scientific technological advances and medical engineering at the nanoscale level [2,3]. All these definitions cover the design and manipulation of nanomaterials. Therefore, nanomaterials, which are one dimension smaller than 100 nanometers, enhance physical, chemical, and biological properties of the original material [4].

Over the last few years, nanotechnology and related disciplines have undergone an exponential growth in applications such as nanomedicine, energy and electronics, and the environment and materials because of the unique properties of nanomaterials [5]. Nanomedicine involves the development of nanoparticles (NPs), among other nanocomponents and devices. For the molecular diagnostics, nanomedicine includes treatment and prevention of human diseases thanks to their compatibility with biomolecules [6].

Currently, NPs have an impact and show an increasing presence in many scientific designs and developments [7]. These have a number of disadvantages, such as the cytotoxic effects in living organisms, which may limit their use within the clinical setting [8]. However, important advantages, which make them an ideal approach for biomedical applications, such as their intrinsic ability to enter the human body through inhalation, the skin and digestive routes depending on their physicochemical properties potentially accesses vital organs through the blood flow [9]. However, some key factors must be taken into consideration during the bio-nano-interface construction: (i) The interaction of nanoparticles with their ecosystem, mainly with other nanomaterials and biomolecules. Some studies show the possibility of using AgNPs as antibacterial agents thanks to high toxicity against human pathogenic bacteria. In this sense, Singh T. et al. have demonstrated the use of endophytic fungi Alternaria sp. to synthetize AgNPs [10]. (ii) Their physicochemical properties achieve a suitable design such as particle size, shape, dispersity, surface charge, and protein corona effects. Protein corona is a complex plasma proteins layer around the NPs that takes place after systemic administration, when nanoparticles are exposed to physiological proximal fluids, which is mostly blood.

The adsorption of dozens of proteins with varying identities and quantities on the NPs can modify their physicochemical identity, cellular uptake, targeting, circulation lifetime in the blood, and influence the physiological response and toxicity [11]. In this sense, a number of molecules can be used to maintain the integrity and stability of NPs in biological fluids [12]. Dutta P.H. et al. synthesized and characterized two types of NPs, AgNPs, and AuNPs, in order to design an antimalarial nanomaterial. For it, they optimized the size, shape, and surface morphology of the bio-synthesized NPs and showed that AgNPs had insignificant and lower cytotoxicity against several human cancer cell lines than AuNPs. [13]. (iii) As well as the interacting bio-compounds (biomolecules, cells, proximal fluids) that favor a physical, chemical, and mechanical relevant process [1,5]. Mirkin showed that siRNA-based gold nanoparticles inhibit its enzymatic degradation and facilitates its uptake by Hela cells [14].

Despite their potential advantages and promising applications, NP entrance in a physiological environment may be problematic due to different intrinsic NP characteristics. These particles may appear embedded in human proximal fluids, inside cells and culture media among others. Thus, there are multiple conditions and a huge variety of biomolecules potentially interacting with the NPs. This previous knowledge is important for predicting their impact [15]. Because of these inherent interactions, the NPs might have a heterogeneous morphology, which is also correlated with the resulting immuno-biocompatibility and safety of these nanomaterials.

In this case, this mini-review focused on the global interactions of NPs and biomolecules in biological environments, which play a critical role in biomedicine applications.

2. Nanoparticle-Cell Dynamics

In general, NPs enter the cells through different internalization mechanisms (Figure 1), accumulate in targeted organs, and are later eliminated.

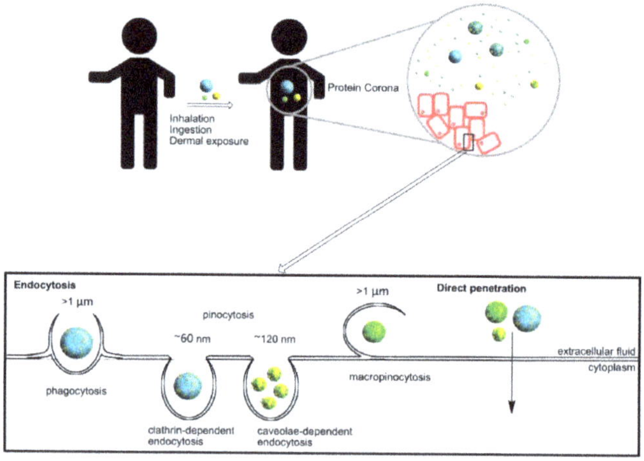

Figure 1. Schematic representation of different ways to enter nanoparticles (NPs) in the human body and inside cells.

Their small size allows them to enter the human body by inhalation, ingestion, or through the skin. Once in the extracellular fluid, they are conjugated with biomolecules presented in the media, which allows them to internalize in the cells.

2.1. Cellular Internalization

Nanomedicines based on NPs must cross the cell membrane by different mechanisms (phagocytosis, micropinocytosis, clathrin-dependent endocytosis, caveolae-dependent endocytosis, or by direct penetration) to produce an effect inside the cells [16] because, usually, cell membranes are impermeable to NP diffusion. Moreover, non-specific internalization mechanisms may induce toxicity. Since a lack of biocompatibility is desirable, it becomes a suitable mechanism for internalization [17].

As expected, the NP size directly affects the internalization process. In fact, NPs in the range of 10–100 nm achieve higher cellular uptake and, on the other hand, small ones imply a higher energy cost to the cells [18]. Usually, NPs larger than 100 nm are internalized by specialized phagocytic cells (such as macrophages, dendritic cells) which allows targeted design.

Furthermore, the optimal size for internalization inside the cells is strongly linked to the NPs surface chemistry. In general, Van der Waals or electrostatic forces are critical in the NP interactions with biomolecules and cells. In fact, several studies show correlation between zeta potential and endocytosis/exocytosis mechanisms [19]. Then, specific cellular internalization could be targeted to favor specific interactions (i.e., employing affinity ligands) as opposed to nonspecific interactions (i.e., hydrophobic). In this sense, antibody-coated NPs present an internalization potential in targeted cells four to eight folds higher than positively or negatively charged NPs without the affinity component [20]. Besides the use of antibodies for targeting delivery, non-specific interactions through chemical moieties are always present and influence target affinities, which must be always taken into consideration.

Protein adsorption also depends on the NP shape and, consequently, affects the cellular uptake. It seems that spherical and highly homogeneous NP conjugates have better cellular uptake than amorphous and non-geometrically symmetric nanoconjugates [21]. Moreover, several authors claim that shape could be employed to prevent non-specific cellular internalization in the targeted cells [19].

2.2. Tumor Accumulation

NPs accumulate preferentially in tumor tissues in comparison with the normal ones [22]. This is mainly because the vessels around the tumoral tissue have a higher permeability (than the normal vessels) and tumors have impaired lymphatic drainage, which leads to retention of the permeated NPs. This is an effect called enhanced permeation and retention (EPR) [23–25].

Tumors are densely packed with cells and the extracellular matrix. Thus, NP size plays an important role in the diffusion and accumulation inside the tumor. The accumulation within the tumors could be modulated by the NP physical dimensions and surface chemistry. In general, diffusion and NP size are inversely correlated [26]. Small size NPs can diffuse freely across tumoral tissue and present a widespread distribution within normal tissues. However, small NPs can easily and quickly clear out. Size is important for other purposes such as when NPs are applied as imaging agents helping to distinguish normal and pathological tissues because they appear only on the tumor periphery thanks to their bigger size.

As previously discussed, the biomolecules adsorption onto the NP surface is directly related with their opsonization and clearance capacity. Therefore, it is related with the blood concentration along with time.

2.3. Elimination

In general, NPs are eliminated from the human body by renal and hepatobiliary routes and need to be done for clinical approval in a reasonable timeframe. Then, drug conjugated NPs must be designed to avoid quick clearance and long period of body maintenance.

As is expected, surface chemistry, shape, and NP size influence elimination. For example, surface chemistry is quite critical in the clearance efficiency -even for small NPs- and polyethylene glycol (PEG) coating promotes more efficient hepatobiliary clearance [27]. Another point is the NP size. The hydrodynamic NP size has a strong influence on the renal clearance, where the glomerular pores are a physical barrier [23,28].

3. Nanoparticle Interactions

Although NPs in biological systems are surrounded by large quantities of biomolecules, depending on the different factors that characterize the biological environment. The NP promotes multiple and different interactions. Multifunctional NPs as nanomedicines (see Figure 2) are embedded in human proximal fluids, inside cells, and inside culture media among others [29]. This implies a huge variety of different microenvironments with additional challenges for the design and development of NPs suitable to be functional in all kinds of conditions. However, depending on the medium conditions like pH [30], ionic strength, oxygen levels, organic matter, etc., NPs present different forms or stages, such as ionized particles [31], which form aggregates or combine into complex aggregates or even interact with other nanomaterials [32]. This is especially relevant because it may be the origin of a heterogeneous morphology, which might be correlated with a lack of stability and immuno-biocompatibility of these nanomaterials [33].

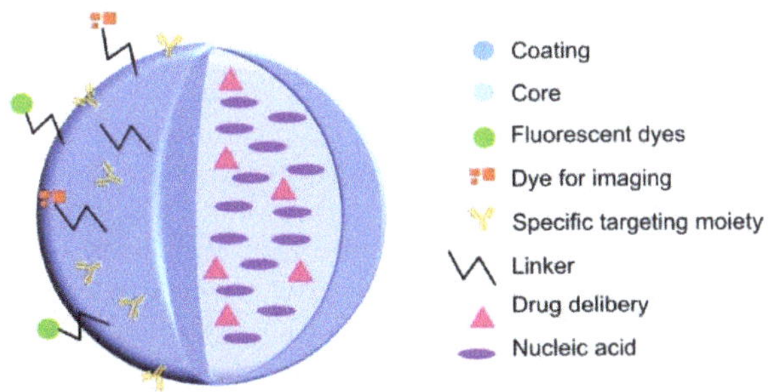

Figure 2. Schematic representation of multi-functional nanoparticles.

NP aggregation and agglomeration have been recognized to affect cellular uptake and even induce potential toxicity based on the nanoparticle composition and the cell type [34]. Aggregation and agglomeration effects are often used in nanotechnology, but both terms are commonly mistaken. Aggregation indicates strongly bonded or fused particles and it occurs when the Van der Waals attractive forces between particles are greater than the electrostatic repulsive forces produced by the nanostructure surface [34,35]. On the other hand, agglomeration indicates more weakly bonded particles and it does not require a definite pattern, shape, and size [35]. Pellegrino F. et at. studied the agglomeration and aggregation influence on the optical properties of TiO_2 NPs demonstrating that this effect can lead to an incorrect assessment of the photoactivity [36]. Zook M.J. et al. [37] developed a bottom-up-based method to produce controllable, reproducible, and stable NP agglomerates in an aqueous medium. They used this method to show how silver NP agglomeration affects hemolytic activity.

The main factors that will determine the type of interactions between NPs are: the complementarity between nanomaterials and their distance and geometry [38]. In addition, it is also essential to know what the main interactions drivers are in an NP assembly. For example, Van der Waals forces form nanocrystal superlattice membranes, electrostatic interactions obtain colloidal dimers, and magnetic interactions where iron oxide NPs coated with azobenzene-terminated catechol ligands self-assemble by UV-light-induced, or even molecular force [38].

An example that demonstrates the importance of the complementarity between the materials and the influence of the forces used in such an interaction is one discussed by Pileni and co-workers. They stress the difference of using octanoic and dodecanoic acids as organic ligands in magnemite NPs in the absence (only with dipolar forces between the magnetic nanoparticles) and the presence of Van der Waals interactions, when the distance is small [39,40].

On the other hand, an interaction between molecules on surfaces is highly dependent on surface functionalization (Figure 3). This implies the presence of reactive chemical moieties on the surface being homo-functional or hetero-functional depending on whether there is only one chemical group on the surface or whether different chemical reactive groups co-exist [41].

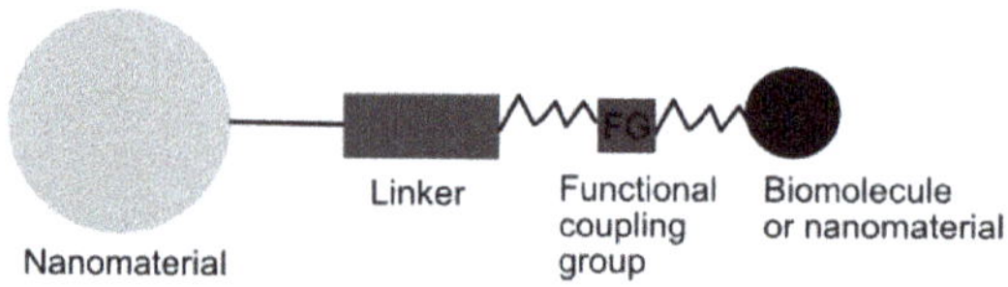

Figure 3. Schematic representation of the strategy to couple nanoparticles and biomolecules or other nanoparticles.

Due to their composition and structure, the surface might not allow different types of interactions. Thus, for example, circulatory cells are covered by a lipid bilayer with proteins and polysaccharides that, depending on the NP exposed groups, will favor one type of interaction mechanism [42]. Another example includes the proteins affected by their molecular weight, charge (greater adsorption near the isoelectric pH), or its stability that influences the number of binding points [43]. A soft protein layer has a low structural stability and a greater number of active centers to interact with, besides other influencing physicochemical factors on the surface (i.e., humectability). The hydrophobicity/hydrophilic surface ratio influences protein reactivity and/or its adsorption properties. Another remarkable feature is the size, including those with a size comparable to that of the NP, which will be more easily adsorbed.

Lastly, it is not only necessary to consider the concentration or size of NPs, but also the species and quantity of resulting products from chemical interactions between NPs.

3.1. Interaction Mechanisms Between Nanoparticles and Biomolecules

There is a wide-open variety of biomolecules, which could interact directly onto the NPs surface or through other biomolecules coating the NPs surface (Figure 2). These layers of coating biomolecules are directly related with the type of organism, biological fluid, cells, etc., among the physicochemical conditions of the media and NP surface, nature, and structure of biomolecules.

According to the literature, the most relevant interacting biomolecules to the NP surfaces are proteins and nucleic acids [44]. Proteins have many different binding sites (as amino acidic key structures and/or post-translational modifications) through specific or non-specific adsorption [43,45]. In addition, the proteins are critical on the immune-biocompatibility of the nanomaterials. Nucleic acids have many different applications as a consequence of its physicochemical stability, mechanical rigidity, easy accessibility, and its high specificity of base pairing, which results in a suitable receptor for molecular nano-construction [46].

Regarding interactions with human biomolecules, two factors must be considered in the description of the interaction [23]. The first one is that NPs in biological systems are surrounded by multiple potentially interacting biomolecules that may modify and saturate their surface. Therefore, custom modified NPs are the ones that may interact specifically with the biomolecules of interest later on. The second factor is NP entering pathways into the human body. This depends on the way it can influence the force of the interaction. For example, NPs entering by inhalation strongly interact with the pulmonary system (proteins and phospholipids).

Two immobilization mechanisms have been studied through an interaction with different types of biomolecules [45]: by simple absorption or by chemical linkages. The immobilization of enzymes on NPs through adsorption is a very useful method because it takes place through non-covalent forces (hydrogen bonding, ionic interactions, and Van der Waal forces), mainly through negatively charged phosphate groups and hydrophobic moieties not disturbing the initial structure of the enzyme or its active site. Immobilization through chemical linkages may lead to the immobilization of biomolecules on a biocompatible matrix, such as within phospholipid bilayers, not interacting with the native structure of the biomolecule and altering its biological activity.

We also find two other types of interaction mechanisms with cells: ligand-receptor interaction and chemical conjugation [47]. An example of the first interaction method is the NP surface functionalization with a receptor, such as streptavidin-biotin. Its non-covalent interaction results in a greater bond strength, which provides resistance to pH, temperature variations, and denaturants. In addition, they have a greater binding affinity to cells. Chemical conjugation simply consists of the coupling of functional groups (such as thiol groups) to the NP surface, which favors subsequent binding to the cell and, in turn, reduces the toxicity of this interaction. A disadvantage of this method is that, in terms of biomedical applications, the covalent binding of the drug to the NP restricts its efficient release, which limits its effectiveness.

3.2. Nanoparticle Design: Influence on Interaction Mechanisms

NPs undergo different changes in a concrete environment such as the generation of a coating protein corona once plasma proteins are adsorbed on its surface. Therefore, it is necessary to study the NP states and characteristics prior to interaction assays [48].

Many NP-based investigations focus on issues affecting NP characteristics and, subsequently, their impact on cellular internalization and biodistribution. Centi J. et al. [49] and Tatini J. et al. [50] talk about the interest of gold nanorods (GNRs) in the biomedical field. GNRs are gold NPs that are elongated along one direction with characteristic optical properties, which depend on the particle size and shape [51]. They are attractive in biomedical optics because of their special and intense absorption band near infrared light (650–1000 nm). Other important features of GNRs include their coating, which are crucial for their biological applications, i.e., conjugation with PEG. In addition, their shape and size are critical for modulating cellular penetration, intracellular localization, and bio-distribution. GNRs may become coated with, which may modify their conformation and cause a loss of their biological activity. Bovine Serum Albumin has been chosen as a protein target to investigate NPs coating with polyethylene glycol (NP-PEG) exposition to biological fluids because it is the most abundant protein in the blood and can transport metal compounds. The Tatini J. et al article proposes CA-125 as the molecular target cancer antigen to model "in vitro" some of the most critical issues that arise from the interactions between GNRs and the bloodstream using an analytical approach.

The physicochemical properties of NPs (Figure 4, some already commented) represent their identity and influence on the synthetic moieties incorporated [52] among all including size, shape, surface, coating and morphology, surface charge, solubility, chemical composition, crystalline structure, and, lastly, the agglomeration status. These properties will also play a characteristic role in relevant mechanisms such as cellular biocompatibility studies.

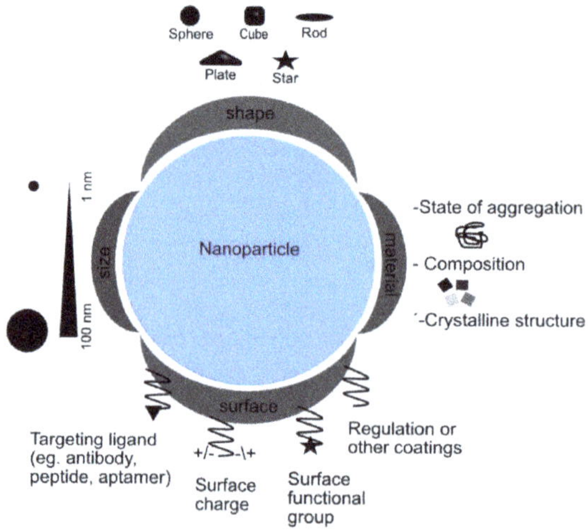

Figure 4. Schematic illustration of the main physicochemical properties of nanoparticles governing interaction mechanisms in biological systems.

3.2.1. Size

Size plays an important role in interactions with the biological system and many biological NP-related mechanisms such as cellular uptake and particle processing efficiency in the endocytic path depending on it [53]. Additionally, the ion release rate, the smaller size, the faster release rate, and the interactions with cell membranes [54]. In general, there is a size-dependent NP toxicity and, therefore, their ability to enter in the human system. As the particle size decreases, the surface/volume ratio increases. Therefore, their contacting surface will increase, which makes penetration into the body easier and increases their toxic effect [54]. NP sizes less than 50 nm through intravenous injection connect to all tissues faster and exert stronger toxic effects [55].

The NP size indicates their "in vivo" distribution, or pharmaceutical behavior [56], and their most direct impact on physiological activity. NP sizes larger than 1 μm cannot easily enter the cell, but they interact with proteins absorbed in the cells. NP sizes greater than 6 nm cannot be excreted by the kindness and accumulate in specific organs [57]. For example, cadmium selenide quantum dots contact stays in the tissue, which causes hepatotoxicity [58].

Sonavane et al. carried out studies on the bio-distribution and bioaccumulation in the blood of gold nanoparticle (AuNP) of different sizes. They observed that smaller ones stayed longer in the bloodstream and accumulated to a greater extent in all organs [59].

3.2.2. Shape

Shape is a physicochemical property that influences the toxicity of materials [60]. NPs have different shapes and structures such as tubes, fibers, spheres, and planes. Therefore, it may also influence their endocytosis process, internalization, bio-distribution, and elimination. For example, spherical nanoparticles of similar size have been found to be easier and faster internalized by endocytosis than rod-shaped nanoparticles, which is explained by a greater membrane wrapping time required for the elongated particles. In addition, the spherical ones are relatively less toxic [21].

3.2.3. Surface Modification

NP-cell interactions and solubility depend on the nature of the NP surface [61]. NP surface coating alteration can modify their magnetic, electrical, chemical, and optical properties, which affects their cytotoxic properties by influencing pharmacokinetics, distribution, accumulation, and toxicity [62].

Surface charges determine the response of the organism to changes in NP shape and size in the form of cellular accumulation, called colloidal behavior [63]. The effect of surface chemistry on NPs affects absorption [64], colloidal behavior, plasma protein binding [65], and crossing the blood-brain barrier [66]. The NP cytotoxicity increased with an increase in surface charge [67]. This suggests that higher positive charges get greater cell electrostatic interactions and, consequently, greater endocytic uptake. However, the uptake of positively charged NPs may produce higher toxicity than negatively charged [68]. NPs with a positively charged surface tended to accumulate more in tumors than negatively charged ones most likely because positively charged density can be more easily separated in the interstitial space and, therefore, internalized by tumor cells [56].

Surface chemical modification is an important strategy utilized in biomedical applications to decreases toxicity, increase stability, and to control and modulate cellular internalization [69]. Surface functionalization is predominantly comprised by polyethylene glycol (PEG), the negative carboxyl group, and neutral groups like hydroxyl group, and amine groups [67]. For example, the NP surface can be functionalized by proper polymers such as PEG to reduce non-specific binding and to get specific binding to cell receptors [70].

Hydrophobicity is another key factor that also affects pharmacokinetics and bio-distribution [70]. NPs with a 2more hydrophobic surface tend to absorb plasma proteins, which reduces the time spent in the bloodstream [71]. A computer molecular simulation study revealed that the surface membrane uptake of hydrophobic C60 agglomerates is thermodynamically favored than semi-hydrophilic ones because of the interior membrane hydrophobicity space in cells [72].

3.2.4. Chemical Composition

NPs chemistry is another fundamental factor contributing to cell interactions. Regarding particle chemistry, Griffitt et al. [73] observed different toxicity in zebrafish, daphnids, and algae species for silver and copper NPs with respect to titanium oxide, which resulted in no toxicity problems.

In addition to these characteristic properties of NPs, their state of aggregation must also be taken into account. Aggregation depends on the surface load, material type, and size, among other factors. It has been shown that higher NP concentrations result in higher aggregation and, consequently, lower toxicity [74]. Accordingly, macrophages remove large particles more efficiently and easily than small ones, which evade this defense mechanism more easily [75].

3.2.5. Protein Corona

Since NPs are injected into the bloodstream, they are exposed to a large amount of biomolecules that form a corona around them [76] (Figure 5). Protein corona is mainly composed of proteins with different affinity interactions: immunoglobulin G, serum albumin, fibrinogen, clusterin, and apolipoproteins [77]. Therefore, NPs experiment changes in their physicochemical properties and their biological identity once the protein corona is formed. Therefore, in order to know the possible adverse effects of the physicochemical, kinetic, dynamic, and thermodynamic interactions of NPs, the characterization of these NP-protein interactions has become one of the main challenges of nanomedicine.

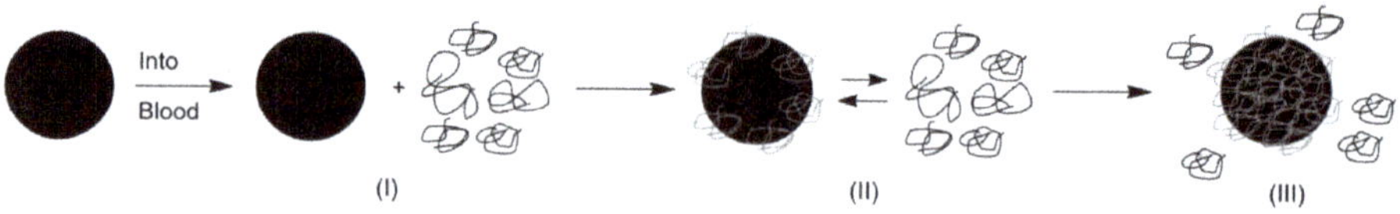

Figure 5. Schematic protein corona formation. First, the introduction of a nanoparticle to fluid/medium enriched in protein content takes place (I). Then, the nanoparticle is coated with proteins, which are abundant and highly mobile (II). Lastly, the protein species are exchanged over time, which results in hard corona of strongly bound proteins (III).

When NPs are incubated in a biological medium, a competitive dynamic process (between soluble biomolecules and surface) take place to form the protein corona. This process is based on the affinity adsorption of proteins on NP surfaces and on protein-protein interactions. According to the Vroman effect [78], the first are bound to NP surface proteins with a high concentration and low affinity and then are gradually replaced by higher affinity proteins present in low concentrations. The protein corona is classified into hard and soft depending on the duration of protein exchanges. Hard corona is a bound layer of proteins with high affinity and long exchange time. Proteins of the hard corona form the closest layer to the NP surface, so they are susceptible to thermodynamically favorable conformational changes (irreversible) depending on the chemistry functionalization, the hydrophobicity or hydrophily, the nature of proximal biological fluid, and the temperature [79]. Soft corona is a low affinity layer of proteins with a fast exchange over time. A recent model [80] suggests that hard corona is bound in a hard way to the NP surface and the soft corona is not directly bound to the NP but with a certain (low) degree of biomolecule interactions. As a result, the protein concentration, particle size, type of nanomaterial, and the surface properties are factors determining the layers of biomolecules and the protein corona density [81].

Depending on the type of administration routes, NPs are subjected to interactions with different kind of biomolecules [82]. The biological environment is another key factor that plays a determinant role in the protein corona formation: the media components, temperature, pH, and the physiological state of the medium. The "in vivo" protein corona formation of biomedical liposomes seems to be more complex than "in vitro" [83]. In consequence, the "in vivo" protein corona characterization is fundamental for biomedical applications.

Different methods and techniques are needed to determine proteins interactions in different biological media because of the large number of proteins at different concentrations that compete to functionalize with the NP surface [84]. Techniques usually described for protein corona evaluation are based on proteomic analysis [80], centrifugation, isothermal calorimetry titration, Ultraviolet and Visible (UV-Visible) spectrometry, Liquid Chromatography with tandem mass spectrometry (LC-MS/MS) quantification, and sodium dodecyl sulfate–polyacrylamide gel (SDS-PAGE) electrophoresis [85].

Therefore, it is essential to understand the relationship between the different properties of nanomaterials and a concrete biological environment in order to understand their stability, viability, behavior, and the results obtained in the different areas of research.

4. Applications of Nanoparticles

In this section, some of the NPs applications are briefly described in order to understand their broad potentials. Although we have focused on nanomedical applications, we should not forget all those other important non-biological applications that have improved the quality of human life (https://www.nanotechproject.org/inventories/) [86].

NPs have attracted great interest for nanobiotechnology applications (Table 1). The design of nanostructures controlling their surface properties is a strategy meant to achieve improved responses aimed at a medical application. Nano-biotechnology plays a central role in nanomedicine and other areas, which aspire to develop highly functional biosensors, molecular switches, and tissue analogs for organs of the body among others. The nano-biotechnological applications to disease treatment, diagnosis, monitoring by bioimaging, biosensing, and drug delivery have been referred to as nanomedicine. Nanomedicine holds significant potential to improve the efficacy of cancer immunotherapy.

Table 1. Main applications of nanoparticles in nanomedicine.

Applications	Findings	Conclusions	References
Tissue and implants engineering	Gold and titanium dioxide nanoparticles have been used to enhance cell proliferation rates for bone and cardiac tissue TiO_2 nanoparticles conjugated with the polymer poly(lactic-co-glycolic acid) (PLGA), decrease harmful effects, match the nanostructured roughness of bone, and improve their cell performance. Nanofibers that serve as a peptide scaffold allow the regeneration of the axonal tissue.	Nanotechnology in tissue engineering is used to create, repair, and/or replace cells, tissues, and organs combining cells with bio-nanomaterials, and to provide the best micro-environment where cells must grow. Nano-scaffolds are used in tissue and implants engineering to regenerate central nervous system cells and possibly other organs.	[87–89]
Antimicrobial vehicules	Silver and titanium dioxide nanoparticles have antimicrobial properties that allow them to be used in surgical mask coatings by eliminating bacteria and viruses.	Drug coated nanoparticles have shown the potential to repel microorganisms and to act as a prevention tool. A unique property of nanomaterials is their high surface-to-volume ratio. Therefore, minuscule amounts of nanoparticles can lend substantial antimicrobial effects.	[90–92]
Gene delivery	Silica nanospheres functionalized with ammonium cation groups allow transfecting cell lipids, polymers, graphene, carbon nanotubes, nanospheres, and different types of inorganic particles to be used.	Nanoparticles have a great potential as vectors to deliver genetic material into living cells.	[93–95]
Cell separation	Magnetic nanoparticles (MNPs) allow magnetic bio-separations with low toxicity and high biocompatibility. At physiological pH and high salt concentrations, nanocomposites acquire a positive charge for easy electrostatic interactions. In general, the magnetic bio-separation of targeted biomolecules occurs thanks to the interaction between MNPs and a targeted molecule with a magnetic force.	Magnetic nanoparticles (MNPs) can be employed to separate biomolecules such as proteins, deoxyribonucleic acid (DNA), cells, bacteria, genes, and viruses depending on the specific functionalization of MNPs.	[95,96]

Table 1. *Cont.*

Applications	Findings	Conclusions	References
Biofuels	The use of Fe (0) nanoparticles favors the activity of bio-hydrogen production under anaerobic conditions.	Nanoparticles are attractive materials to produce sustainable energy resources, mainly biofuels, thanks to their large surface/volume ratio, which provides a greater number of active sites where they catalyze bio-hydrogen, biogas, biodiesel, and bioethanol production in a high yield.	[97–99]
Drug Delivery System (DDS)	A platinum derivate of a bile acid conjugated with multifunctional polymer-coated bio-ferrofluids as anti-tumor agent in osteosarcoma (MG-63) and T-cell leukemic (Jurkat) cells. The use of gold nanoparticles, polymer nanoparticles, or liposomes, among others, as excellent tumor peptide vaccine carriers play an important role in anti-tumor immunotherapy.	Nanoparticles-based drug delivery system (DDS) have been in the core of attention due to their unique and superior properties. These systems can enhance therapeutic efficacy by producing more favorable bio-availability, serum stability, and pharmacokinetics. Nanoparticle formulations provide better penetration and allow slow and controlled release of drug molecules at the target site for bioactivity	[100–103]
Anti-cancer chemotherapy	Chemical analogues with platinum (II)-based drugs or ruthenium-based antimetastatic agents have anti-cancer properties. The behavior and the biological properties of novel gold compounds containing different ligands have been reported for human ovarian cancer cells. One of the most studied gold (III) compounds is Auranofinan orally effective anti-rheumatic administered drug and an anti-cancer treatment.	Nanoparticles technology offers a series of advantages for drug delivery such as high loading yield, combination therapy, controlled release, prolonged circulation, and targeted delivery. Recently, platinum (II), ruthenium, and gold (III) compounds-based anti-cancer chemotherapy has been reported to kill cancer cells. Most of these studies have been done using proteomics approaches.	[49,104–115]

Table 1. *Cont.*

Applications	Findings	Conclusions	References
Biosensors	An enzyme-linked immunosorbent assay (ELISA) was developed in which nanoparticles (AuNPs) were used as carriers of the signalling antibody, anti-CA15-3-HRP, for the analysis of CA15-3, which is an important tumour marker useful for the follow-up of breast cancer. The use of magnetic nanoparticles as proximity sensors in magnetic resonance (NMR) is known as diagnostic magnetic resonance (DMR).	AuNPs can be used to improve the performance of studies, such as the classical ELISA test, which achieves greater sensitivities. The idea of using nanomaterials in biosensors arose from the possibility of lowering the detection limit (LOD) and improving the signal-to-noise ratio. A diagnostic magnetic resonance (DMR) is a powerful biosensor technology that offers advantages over other detection techniques as well as broad applicability for profiling different types of targets (DNA, proteins, metabolites, and cells).	[116–119]

Nanomedical Applications: Immunotherapy

Immunotherapy has become one of the effective treatment modalities for cancer: cytokine therapy, checkpoint-blockade therapy, adoptive T-cell transfer, and Chimeric Antigen Receptor T(CAR-T) cell therapy [120]. Immunotherapy not only treats primary tumors but also prevents metastasis and recurrence. Another opportunity for combinatorial immunotherapy is based on NP platforms because of their improved methods for tumor-cell detection, tumor imaging, and their ability to efficiently deliver drugs to target sites and protect drugs from endogenous enzymes [121]. Therefore, it is relevant to highlight how NPs may be engineered to overcome immunotherapy obstacles. In this mini review, we have discussed how NPs properties affect a biological mechanism and how they influence cellular internalization, biodistribution, and elimination. Therefore, we have enough information to understand how they alter immune responses.

NPs can release agents in response to biochemical changes in the target micro-environment (pH, redox potential, and enzymes) or to external stimuli (light, electrical, and magnetic fields) [122]. Due to that, targeted delivery of NPs and controlled drug release may allow the activation of immunotherapies in the action sites [123]. The use of NPs for delivery antigens, adjuvants, and other therapeutic agents resulted in more specific targeting and a better outcome in contrast to conventional immunotherapy. Advanced biomaterials and drug delivery systems, such as NPs and the use of T cells, have been designed to improve immunotherapy [124]. Moreover, NPs can deliver cytotoxic agents to tumor cells killing most of all the target cells with low concentrations of immune-stimulating drugs thanks to their potential to amplify T cell responses [120].

NP physicochemical properties can be tuned to stimulate the innate immune cells and to promote NP-immune cell interactions, which is a good therapeutic option [125]. Different strategies to enhance the efficacy of NP immunotherapy are the following [126,127]: Controlling the hydrophobicity surface (using hydrophilic polymers such as PEG) and a shape and rigidity optimization of NPs must reduce nonspecific uptake, which results in an efficient internalization. Enhancing tissue and cell penetration has been possible using peptide and chemical modifications to the NP surface such as cyclic iRGD peptide (CRGDK/RGDP/EC). Another important factor is targeting NPs and their bio-distribution to immune cells with ligands on NPs, such as T lymphocyte or B lymphocyte targeting. Nano-sized NPs have the advantage of accumulating within the tumor microenvironment with specific targeting, which minimizes off-target toxicity [125]. Then, once NPs reach the target cell, their biological activity occurs when they travel to the suitable intracellular compartment [127]. As a result, cationic polymers, pH-sensitive biomaterials, virus-derived cell-penetrating peptides, and direct cytosolic delivery must be used on NP in order to conduct appropriate intracellular delivery of NPs. Lastly, another approach to control immunotherapy is controlling the release kinetic.

The most common nanocarriers allowing specificity are liposomes, micelles, dendrimers, gold NPs, iron oxide NPs, carbon NPs, and quantum dots (NPs for tumor immunotherapy). Liposomes [128] are highly biocompatible and can be functionalized. However, they are widely studied for cancer immunotherapy. Micelles have a range application in cancer treatment because of their biodegradability and nontoxicity formulations, which makes them suitable for carrying therapeutic payloads. In addition, dendrimers [129] offer a highly specific NP physical properties thanks to their stepwise branching synthesis. Inorganic nanoparticles are well studied, such as gold NPs [123]. AuNPs are bio-inert and non-toxic nanocarriers, which, depending on their size, charge, shape, and functional group, may contribute to the efficacy in accumulating different immune cells [125]. The most studied functionalization for cancer immunotherapy is nanoparticles based on poly(lactic-co-glycolic acid) (PLGA NPs) because of their acceptance and biodegradability. Rosalia et al. [130] studied PLGA NPs functionalized with a αCD40-monoclonal antibody agonistic vaccine targeting dendritic cells (DCs). Two different adjuvants targeting the toll like receptor (TLR) were encapsulated into PLGA NPs to induce potent CD8+ T cell responses. In vivo experiments in murine melanoma-OVA mouse model indicated that active targeting of DCs and vaccine delivery resulted in efficient priming of CD8+ T cells, tumor control, and prolonged survival of the tumor-bearing mice.

Several programs work in integrated and interconnected research focused on therapeutically modifying the tumour micro-environment, (re)activation of anti-cancer immunity, and corresponding Drug Delivery System (DDS) [131]. Initially, they aim to develop new tumor-targeted drugs to selectively block key innate and adaptive immune checkpoints, such as PD-1, TIM-3, and CD47, in the tumor micro-environment. Furthermore, they aim to develop new tumor-targeted drugs to selectively activate key co-stimulatory receptors of the tumor necrosis factor receptor superfamily (TNFRSF) in the tumor micro-environment.

On the other hand, suitable drug delivery systems (DDS) could be developed using modern drug formulations based on nanotechnology and surface chemistry to achieve tumor-localized release and optimal localized co-stimulation of anti-cancer immunity. These developments will be attended by label-free detection of protein interactions by means of advanced bioanalysis methods [132]. That could ensure induction and execution of anti-cancer immune responses in the absence of systemic immune-related side-effects.

Lastly, immunotherapies help to amplify the knowledge and manipulation of the immune system and nanotechnology may be the cause of engineering remarkable mechanisms to produce an effective and long-lasting immune response against cancer.

5. Conclusions and Perspectives

Nanotechnology has significantly impacted medicine. In the past decade, studies about the biological response to NPs are greatly investigated in parallel with nano-bio interactions, which have influenced NP design. Both were in concordance with the evolution of NPs for biomedical applications. Many studies have investigated and demonstrated that NPs can enter into the human system. Therefore, the NP characteristics on biological systems, such as their physicochemical properties (size, shape, surface, coating and morphology, surface charge, hydrophobicity, chemical composition, structure, and the state of agglomeration), the types of biomolecules present, and the bio-identity of NP protein corona are important issues to characterize in order to know how they interact with cells, organisms, biological medium, biomolecules, and other biological systems or even with other nanomaterials. These studies helped determine their possible biocompatibility and toxicity in biological micro-environments and to engineer nontoxic nanomaterials, which may be used in biomedical applications.

With the potentially wide application of NPs in the future, these may be extensively used in various fields, especially in immunotherapy for clinical diagnosis and therapy based on their size, biocompatibility, surface chemistry, and adjustable toxicity. Immunotherapy combined with nanomedicines may be used to treat different types of cancer due to their excellent efficacy in penetration, specific retention, and killing of tumor cells.

The human proteome study [133] can be an arduous and discouraging task due to the high number of proteins, encoded by around 25,000 different genes, from which multiple protein variants are generated by post-translational modifications. The concept of proteomics involves a comprehensive study on the structures, localizations, post-translational modifications, functions, and interactions of all proteins expressed by an organism at a certain time and under certain conditions. The nanotechnology field has been expanded by providing innovative methods capable of responding to proteomic demands. In this sense, nanotechnology applications in proteomics have established a novel technical platform termed "nanoproteomics." Detection techniques without labels are useful in the study of protein interaction kinetics, thanks to avoiding steric impediments caused by the presence of labels. The design and development of new multi-functional platforms based on nanomedicine could be of great interest in the unlabeled detection of protein-protein interactions given the possibility of synthesizing de novo proteins "in vitro" in the presence of these nano-systems.

In conclusion, the progress in nano-bio studies can potentially improve nano-medical applications and ensure a sustainable future.

Author Contributions: C.A.-S., R.M.-R., M.F. writing original draft preparation. C.A.-S., T.N., A.L.-V., P.J.-V., H.F.-G., M.J.A.P., R.G., V.A.-F., R.M.-R., M.F. writing—review and editing. C.A.-S., T.N., A.L.-V., P.J.-V., H.F.-G., V.A.-F. designing figures. Conceptualization, R.G., M.J.A.P., R.M.-R. & M.F. The manuscript has been approved by all the co-authors.

Funding: We gratefully acknowledge financial support from the Spanish Health Institute Carlos III (ISCIII) for the grants: FIS PI17/01930 and CB16/12/00400. Fundación Solórzano FS/38-2017. The Proteomics Unit belongs to ProteoRed, PRB3-ISCIII, supported by grant PT17/0019/0023, of the PE I + D + I 2017-2020, funded by ISCIII and FEDER.

Acknowledgments: We gratefully acknowledge financial support from the Spanish Health Institute Carlos III (ISCIII) for the grants: FIS PI17/01930 and CB16/12/00400. Fundación Solórzano FS/38-2017. The Proteomics Unit belongs to ProteoRed, PRB3-ISCIII, supported by grant PT17/0019/0023, of the PE I + D + I 2017-2020, funded by ISCIII and FEDER. A. Landeira-Viñuela is supported by VIII Centenario USAL-BancoSantander PhD Grant. We also thank G.R.G. for language edition services.

Conflicts of Interest: The authors declare no conflict of interest.

References

1. Dean, S.; Mansoori, G.; Fauzi Soelaiman, T. Nanotechnology—An Introduction for the Standards Community. *J. ASTM Int.* **2005**, *2*, 13110. [CrossRef]
2. Schmidt, K.F. *Nanofrontiers: Visions for the Future of Nanotechnology*; Woodrow Wilson International Center for Scholars: Washington, DC, USA, 2007.
3. National Nanotechnology Initiative. Available online: http://www.nano.gov (accessed on 14 June 2019).
4. Davies, J.C. *Nanotechnology Oversight*; Project on Emerging Nanotechnologies: Washington, DC, USA, 2008.
5. Bera, A.; Belhaj, H. Application of Nanotechnology by Means of Nanoparticles and Nanodispersions in Oil Recovery—A Comprehensive Review. *J. Nat. Gas Sci. Eng.* **2016**, *34*, 1284–1309. [CrossRef]
6. Riehemann, K.; Schneider, S.W.; Luger, T.A.; Godin, B.; Ferrari, M.; Fuchs, H. Nanomedicine-Challenge and Perspectives. *Angew. Chem. Int. Ed.* **2009**, *48*, 872–897. [CrossRef] [PubMed]
7. Sanchez, F.; Sobolev, K. Nanotechnology in Concrete-A Review. *Constr. Build. Mater.* **2010**, *24*, 2060–2071. [CrossRef]
8. Ingle, A.P.; Duran, N.; Rai, M. Bioactivity, Mechanism of Action, and Cytotoxicity of Copper-Based Nanoparticles: A Review. *Appl. Microbiol. Biotechnol.* **2014**, *98*, 1001–1009. [CrossRef] [PubMed]
9. Verma, A.; Stellacci, F. Effect of Surface Properties on Nanoparticle-Cell Interactions. *Small* **2010**, *6*, 12–21. [CrossRef] [PubMed]

10. Singh, T.; Jyoti, K.; Patnaik, A.; Singh, A.; Chauhan, R.; Chandel, S.S. Biosynthesis, Characterization and Antibacterial Activity of Silver Nanoparticles Using an Endophytic Fungal Supernatant of Raphanus Sativus. *J. Genet. Eng. Biotechnol.* **2017**, *15*, 31–39. [CrossRef] [PubMed]
11. Phogat, N.; Kohl, M.; Uddin, I.; Jahan, A. Interaction of Nanoparticles With Biomolecules, Protein, Enzymes, and Its Applications. In *Precision Medicine*; Elsevier: Atlanta, GA, USA, 2018; pp. 253–276.
12. Guerrini, L.; Alvarez-Puebla, R.A.; Pazos-Perez, N. Surface Modifications of Nanoparticles for Stability in Biological Fluids. *Materials* **2018**, *11*, 1154. [CrossRef]
13. Dutta, P.P.; Bordoloi, M.; Gogoi, K.; Roy, S.; Narzary, B.; Bhattacharyya, D.R.; Mohapatra, P.K.; Mazumder, B. Antimalarial Silver and Gold Nanoparticles: Green Synthesis, Characterization and in Vitro Study. *Biomed. Pharmacother.* **2017**, *91*, 567–580. [CrossRef]
14. Vigneshwaran, N.; Jain, P. Biomolecules–Nanoparticles: Interaction in Nanoscale. In *Metal Nanoparticles in Microbiology*; Rai, M., Duran, N., Eds.; Springer: Berlin/Heidelberg, Germany, 2011; pp. 135–150.
15. Byrne, H.J.; Ahluwalia, A.; Boraschi, D.; Fadeel, B.; Gehr, P.; Gutleb, A.C.; Kendall, M.; Papadopoulos, M.G. The bio-nano-interface in predicting nanoparticle fate and behaviour in living organisms: Towards grouping and categorising nanomaterials and ensuring nanosafety by design. *BioNanoMaterials* **2013**, *14*, 195–216.
16. Yameen, B.; Choi, W.I.; Vilos, C.; Swami, A.; Shi, J.; Farokhzad, O.C. Insight into Nanoparticle Cellular Uptake and Intracellular Targeting. *J. Control. Release* **2014**, *190*, 485–499. [CrossRef] [PubMed]
17. Hu, G.; Jiao, B.; Shi, X.; Valle, R.P.; Fan, Q.; Zuo, Y.Y. Physicochemical Properties of Nanoparticles Regulate Translocation across Pulmonary Surfactant Monolayer and Formation of Lipoprotein Corona. *ACS Nano* **2013**, *7*, 10525–10533. [CrossRef] [PubMed]
18. Shang, L.; Nienhaus, K.; Nienhaus, G. Engineered Nanoparticles Interacting with Cells: Size Matters. *J. Nanobiotechnol.* **2014**, *12*, 5. [CrossRef] [PubMed]
19. Zhao, J.; Stenzel, M.H. Entry of Nanoparticles into Cells: The Importance of Nanoparticle Properties. *Polym. Chem.* **2018**, *9*, 259–272. [CrossRef]
20. Wilhelm, C.; Billotey, C.; Roger, J.; Pons, J.N.; Bacri, J.; Gazeau, F. Intracellular Uptake of Anionic Superparamagnetic Nanoparticles as a Function of Their Surface Coating. *Biomaterials* **2003**, *24*, 1001–1011. [CrossRef]
21. Champion, J.A.; Mitragotri, S. Role of Target Geometry in Phagocytosis. *Proc. Natl. Acad. Sci. USA* **2006**, *103*, 4930–4934. [CrossRef] [PubMed]
22. Maeda, H. Toward a Full Understanding of the EPR Effect in Primary and Metastatic Tumors as Well as Issues Related to Its Heterogeneity. *Adv. Drug Deliv. Rev.* **2015**, *91*, 3–6. [CrossRef]
23. Buzea, C.; Pacheco, I.I.; Robbie, K. Nanomaterials and Nanoparticles: Sources and Toxicity. *Biointerphases* **2007**, *2*, MR17–MR71. [CrossRef]
24. Pirollo, K.F.; Chang, E.H. Does a Targeting Ligand Influence Nanoparticle Tumor Localization or Uptake? *Trends Biotechnol.* **2008**, *26*, 552–558. [CrossRef]
25. Maeda, H.; Wu, J.; Sawa, T.; Matsumura, Y.; Hori, K. Tumor Vascular Permeability and the EPR Effect in Macromolecular Therapeutics: A Review. *J. Control. Release* **2000**, *65*, 271–284. [CrossRef]
26. Ernsting, M.J.; Murakami, M.; Roy, A.; Li, S.-D. Factors Controlling the Pharmacokinetics, Biodistribution and Intratumoral Penetration of Nanoparticles. *J. Control. Release* **2013**, *172*, 782–794. [CrossRef] [PubMed]
27. Otsuka, H.; Nagasaki, Y.; Kataoka, K. PEGylated Nanoparticles for Biological and Pharmaceutical Applications. *Adv. Drug Deliv. Rev.* **2003**, *55*, 403–419. [CrossRef]
28. Boczkowski, J.; Lanone, S. Respiratory Toxicities of Nanomaterials—A Focus on Carbon Nanotubes. *Adv. Drug Deliv. Rev.* **2012**, *64*, 1694–1699. [CrossRef] [PubMed]
29. Mu, Q.; Jiang, G.; Chen, L.; Zhou, H.; Fourches, D.; Tropsha, A.; Yan, B. Chemical Basis of Interactions Between Engineered Nanoparticles and Biological Systems. *Chem. Rev.* **2014**, *114*, 7740–7781. [CrossRef] [PubMed]
30. Lau, C.P.; Abdul-Wahab, M.F.; Jaafar, J.; Chan, G.F.; Rashid, N.A.A. Effect of PH and Biological Media on Polyvinylpyrrolidone-Capped Silver Nanoparticles. *AIP Conf. Proc.* **2016**, *1756*, 1–8.
31. Carrillo-Carrión, C.; Nazarenus, M.; Paradinas, S.S.; Carregal-Romero, S.; Almendral, M.J.; Fuentes, M.; Pelaz, B.; del Pino, P.; Hussain, I.; Clift, M.J.; et al. Metal Ions in the Context of Nanoparticles toward Biological Applications. *Curr. Opin. Chem. Eng.* **2014**, *4*, 88–96. [CrossRef]

32. Sharma, V.K.; Sayes, C.M.; Guo, B.; Pillai, S.; Parsons, J.G.; Wang, C.; Yan, B.; Ma, X. Interactions between Silver Nanoparticles and Other Metal Nanoparticles under Environmentally Relevant Conditions: A Review. *Sci. Total Environ.* **2019**, *653*, 1042–1051. [CrossRef] [PubMed]
33. Peer, D. Immunotoxicity Derived from Manipulating Leukocytes with Lipid-Based Nanoparticles. *Adv. Drug Deliv. Rev.* **2012**, *64*, 1738–1748. [CrossRef]
34. Albaneset, A.; Chang, W.C. Effect of gold nanoparticle aggregation on cell uptake and toxicity. *ACS Nano* **2011**, *57*, 5478–5489. [CrossRef]
35. Nichols, G.; Byard, S.; Bloxham, M.J.; Botterill, J.; Dawson, N.J.; Dennis, A.; Diart, V.; North, N.C.; Sherwood, J.D. A Review of the Terms Agglomerate and Aggregate with a Recommendation for Nomenclature Used in Powder and Particle Characterization. *J. Pharm. Sci.* **2002**, *91*, 2103–2109. [CrossRef]
36. Pellegrino, F.; Pellutiè, L.; Sordello, F.; Minero, C.; Ortel, E.; Hodoroaba, V.D.; Maurino, V. Influence of Agglomeration and Aggregation on the Photocatalytic Activity of TiO_2 Nanoparticles. *Appl. Catal. B Environ.* **2017**, *216*, 80–87. [CrossRef]
37. Zook, J.M.; MacCuspie, R.I.; Locascio, L.E.; Halter, M.D.; Elliott, J.T. Stable Nanoparticle Aggregates/Agglomerates of Different Sizes and the Effect of Their Size on Hemolytic Cytotoxicity. *Nanotoxicology* **2011**, *5*, 517–530. [CrossRef] [PubMed]
38. Yan, C.; Wang, T. A New View for Nanoparticle Assemblies: From Crystalline to Binary Cooperative Complementarity. *Chem. Soc. Rev.* **2017**, *46*, 1483–1509. [CrossRef] [PubMed]
39. Singamaneni, S.; Bliznyuk, V.N.; Binek, C.; Tsymbal, E.Y. Magnetic Nanoparticles: Recent Advances in Synthesis, Self-Assembly and Applications. *J. Mater. Chem.* **2011**, *21*, 16819. [CrossRef]
40. Mørup, S.; Hansen, M.F.; Frandsen, C. Magnetic Interactions between Nanoparticles. *Beilstein J. Nanotechnol.* **2010**, *1*, 182–190. [CrossRef] [PubMed]
41. Dyawanapelly, S.; Mehrotra, P.; Ghosh, G.; Jagtap, D.D.; Dandekar, P.; Jain, R. How the Surface Functionalized Nanoparticles Affect Conformation and Activity of Proteins: Exploring through Protein-Nanoparticle Interactions. *Bioorg. Chem.* **2019**, *82*, 17–25. [CrossRef] [PubMed]
42. Gehr, P. Interaction of Nanoparticles with Biological Systems. *Colloids Surf. B Biointerfaces* **2018**, *172*, 395–399. [CrossRef] [PubMed]
43. Firkowska-Boden, I.; Zhang, X.; Jandt, K.D. Controlling Protein Adsorption through Nanostructured Polymeric Surfaces. *Adv. Healthc. Mater.* **2018**, *7*, 1700995. [CrossRef] [PubMed]
44. Hemmelmann, P. Nanoparticles, Proteins, and Nucleic Acids: Biotechnology Meets Materials Science. *Angew. Chem. Int. Ed.* **2001**, *40*, 4128–4158.
45. Arsalan, A.; Younus, H. Enzymes and Nanoparticles: Modulation of Enzymatic Activity via Nanoparticles. *Int. J. Biol. Macromol.* **2018**, *118*, 1833–1847. [CrossRef]
46. Saallah, S.; Lenggoro, I.W. Nanoparticles Carrying Biological Molecules: Recent Advances and Applications. *KONA Powder Part. J.* **2018**, *2018*, 89–111. [CrossRef]
47. Singh, B.; Mitragotri, S. Harnessing Cells to Deliver Nanoparticle Drugs to Treat Cancer. *Biotechnol. Adv.* **2019**. [CrossRef] [PubMed]
48. Zhao, Z.; Ukidve, A.; Krishnan, V.; Mitragotri, S. Effect of Physicochemical and Surface Properties on in Vivo Fate of Drug Nanocarriers. *Adv. Drug Deliv. Rev.* **2019**, *143*, 3–21. [CrossRef] [PubMed]
49. Centi, S.; Tatini, F.; Ratto, F.; Gnerucci, A.; Mercatelli, R.; Romano, G.; Landini, I.; Nobili, S.; Ravalli, A.; Marrazza, G.; et al. In Vitro Assessment of Antibody-Conjugated Gold Nanorods for Systemic Injections. *J. Nanobiotechnol.* **2014**, *12*, 55. [CrossRef] [PubMed]
50. Tatini, F.; National, I.; Massai, L. Size Dependent Biological Profiles of PEGylated Gold Nanorods. *J. Mater. Chem. B* **2014**, *36*, 6072–6080. [CrossRef]
51. Pérez-Juste, J.; Pastoriza-Santos, I.; Liz-Marzán, L.M.; Mulvaney, P. Gold Nanorods: Synthesis, Characterization and Applications. *Coord. Chem. Rev.* **2005**, *249*, 1870–1901. [CrossRef]
52. Khan, I.; Saeed, K.; Khan, I. Nanoparticles: Properties, Applications and Toxicities. *Arab. J. Chem.* **2017**. [CrossRef]
53. Aillon, K.L.; Xie, Y.; El-Gendy, N.; Berkland, C.J.; Forrest, M.L. Effects of Nanomaterial Physicochemical Properties on in Vivo Toxicity. *Adv. Drug Deliv. Rev.* **2009**, *61*, 457–466. [CrossRef] [PubMed]
54. Huang, Y. The Toxicity of Nanoparticles Depends on Multiple Molecular and Physicochemical Mechanisms. *Int. J. Mol. Sci.* **2019**, *18*, 2702. [CrossRef] [PubMed]

55. Id, M.A.; Moosavi, M.A.; Rahmati, M.; Falahati, M. Health Concerns of Various Nanoparticles: A Review of Their in Vitro and in Vivo Toxicity. *Nanomaterials* **2018**, *8*, 634.
56. Hoshyar, N.; Gray, S.; Han, H.; Bao, G. The Effect of Nanoparticle Size on in Vivo Pharmacokinetics and Cellular Interaction. *Nanomedicine* **2016**, *11*, 673–692. [CrossRef] [PubMed]
57. Albanese, A.; Albanese, A.; Tang, P.S.; Chan, W.C.W. The Effect of Nanoparticle Size, Shape, and Surface Chemistry on Biological Systems. *Annu. Rev. Biomed. Eng.* **2014**, *14*, 1–16. [CrossRef] [PubMed]
58. Ballou, B.; Lagerholm, B.C.; Ernst, L.A.; Bruchez, M.P.; Waggoner, A.S. Noninvasive Imaging of Quantum Dots in Mice. *Bioconjug. Chem.* **2004**, *15*, 79–86. [CrossRef] [PubMed]
59. Sonavane, G.; Tomoda, K.; Makino, K. Biodistribution of Colloidal Gold Nanoparticles after Intravenous Administration: Effect of Particle Size. *Colloids Surf. B Biointerfaces* **2008**, *66*, 274–280. [CrossRef] [PubMed]
60. Lee, Y.J.; Ahn, E.; Park, Y. Shape-Dependent Cytotoxicity and Cellular Uptake of Gold Nanoparticles Synthesized Using Green Tea Extract. *Nanoscale Res. Lett.* **2019**, *14*, 129. [CrossRef] [PubMed]
61. Risom, L.; Møller, P.; Loft, S. Oxidative Stress-Induced DNA Damage by Particulate Air Pollution. *Mutat. Res. Fundam. Mol. Mech. Mutagen.* **2005**, *592*, 119–137. [CrossRef] [PubMed]
62. Ospina, S.P.; Favi, P.M.; Gao, M.; Johana, L.; Morales, M.; Pavon, J.J.; Webster, T.J. Shape and Surface Effects on the Cytotoxicity of Nanoparticles: Gold Nanospheres versus Gold Nanostars. *J. Biomed. Mater. Res. Part A* **2015**, *103*, 3449–3462.
63. Curtis, A.C.; Toghani, D.; Wong, B.; Nance, E. Colloidal stability as a determinant of nanoparticle behavior in the brain. *Colloids Surf. B Biointerfaces* **2018**, *170*, 673–682. [CrossRef]
64. Hoshino, A.; Fujioka, K.; Oku, T.; Suga, M.; Sasaki, Y.F.; Ohta, T.; Yasuhara, M.; Suzuki, K.; Yamamoto, K. Physicochemical Properties and Cellular Toxicity of Nanocrystal Quantum Dots Depend on Their Surface Modification. *Nano Lett.* **2004**, *4*, 2163–2169. [CrossRef]
65. Pietroiusti, A.; Massimiani, M.; Fenoglio, I.; Colonna, M.; Valentini, F.; Palleschi, G.; Camaioni, A.; Magrini, A.; Siracusa, G.; Bergamaschi, A.; et al. Low Doses of Pristine and Oxidized Single-Wall Carbon Nanotubes Affect Mammalian Embryonic Development. *ACS Nano* **2011**, *5*, 4624–4633. [CrossRef]
66. Georgieva, J.V.; Kalicharan, D.; Couraud, P.; Romero, I.A.; Weksler, B.; Hoekstra, D.; Zuhorn, I.S. Surface Characteristics of Nanoparticles Determine Their Intracellular Fate in and Processing by Human Blood—Brain Barrier Endothelial Cells In Vitro. *Mol. Ther.* **2009**, *19*, 318–325. [CrossRef] [PubMed]
67. Foroozandeh, P.; Aziz, A.A. Insight into Cellular Uptake and Intracellular Trafficking of Nanoparticles. *Nanoscale Res. Lett.* **2018**, *13*, 339. [CrossRef] [PubMed]
68. Hu, D.; Kantner, X.K.; Geidel, X.C.; Brandholt, S.; De Cock, I.; Soenen, S.J.H.; Gil, P.R.; Montenegro, J.; Braeckmans, K.; Nienhaus, G.U.; et al. Polymer-Coated Nanoparticles Interacting with Proteins and Cells: Focusing on the Sign of the Net Charge. *ACS Nano* **2013**, *7*, 3253–3263.
69. Chompoosor, A.; Saha, K.; Ghosh, P.S.; Macarthy, D.J.; Miranda, O.R.; Zhu, Z.-J.; Arcaro, K.F.; Rotello, V.M. The Role of Surface Functionality on Acute Cytotoxicity, ROS Generation and DNA Damage by Cationic Gold Nanoparticles. *Small* **2010**, *6*, 2246–2249. [CrossRef] [PubMed]
70. Mahmoudi, M.; Lynch, I.; Ejtehadi, M.R.; Monopoli, M.P.; Bombelli, F.B.; Laurent, S. Protein À Nanoparticle Interactions: Opportunities and Challenges. *Chem. Rev.* **2011**, *111*, 5610–5637. [CrossRef] [PubMed]
71. Duan, X.; Li, Y. Physicochemical Characteristics of Nanoparticles Affect Circulation, Biodistribution, Cellular Internalization, and Trafficking. *Small* **2013**, *9*, 1521–1532. [CrossRef] [PubMed]
72. Li, Y.; Chen, X.; Gu, N. Computational Investigation of Interaction between Nanoparticles and Membranes: Hydrophobic/Hydrophilic Effect. *J. Phys. Chem. B* **2008**, *112*, 16647–16653. [CrossRef] [PubMed]
73. Griffitt, R.J.; Luo, J.; Gao, J.; Bonzongo, J.C.; Barber, D.S. Effects of Particle Composition and Species on Toxicity of Metallic Nanomaterials in Aquatic Organisms. *Environ. Toxicol. Chem.* **2008**, *27*, 1972–1978. [CrossRef]
74. Gatoo, M.A.; Naseem, S.; Arfat, M.Y.; Mahmood Dar, A.; Qasim, K.; Zubair, S. Physicochemical Properties of Nanomaterials: Implication in Associated Toxic Manifestations. *BioMed Res. Int.* **2014**, *2014*, 498420. [CrossRef]
75. Van Der Zande, M.; Vandebriel, R.J.; Van Doren, E.; Kramer, E.; Herrera Rivera, Z.; Serrano-Rojero, C.S.; Gremmer, E.R.; Mast, J.; Peters, R.J.B.; Hollman, P.C.H.; et al. Distribution, Elimination, and Toxicity of Silver Nanoparticles and Silver Ions in Rats after 28-Day Oral Exposure. *ACS Nano* **2012**, *6*, 7427–7442. [CrossRef]

76. Peng, Q.; Zhang, S.; Yang, Q.; Zhang, T.; Wei, X.Q.; Jiang, L.; Zhang, C.L.; Chen, Q.M.; Zhang, Z.R.; Lin, Y.F. Preformed Albumin Corona, a Protective Coating for Nanoparticles Based Drug Delivery System. *Biomaterials* **2013**, *34*, 8521–8530. [CrossRef] [PubMed]
77. Choi, E.; Webster, T.J.; Kim, S. Effect of the Protein Corona on Nanoparticles for Modulating Cytotoxicity and Immunotoxicity. *Int. J. Nanomed.* **2015**, *10*, 97.
78. Hirsh, S.L.; Mckenzie, D.R.; Nosworthy, N.J.; Denman, J.A.; Sezerman, O.U.; Bilek, M.M.M. Colloids and Surfaces B: Biointerfaces The Vroman Effect: Competitive Protein Exchange with Dynamic Multilayer Protein Aggregates. *Colloids Surf. B Biointerfaces* **2013**, *103*, 395–404. [CrossRef] [PubMed]
79. Megido, L.; Díez, P.; Fuentes, M. Nanoproteomics approaches in biomarcker Discovery: The critical role of protein corona on nanoparticles as drug carriers. In *Nanotechnologies in Preventive and Regenerative Medicine*; Uskokovic, V., Uskokovic, P.D., Eds.; Elsevier: Cambridge, UK, 2018; pp. 225–239.
80. Liu, W.; Rose, J.; Plantevin, S.; Auffan, M.; Bottero, J.Y.; Vidaud, C. Protein Corona Formation for Nanomaterials and Proteins of a Similar Size: Hard or Soft Corona? *Nanoscale* **2013**, *5*, 1658–1665. [CrossRef] [PubMed]
81. Xiao, W.; Gao, H. The Impact of Protein Corona on the Behavior and Targeting Capability of Nanoparticle-Based Delivery System. *Int. J. Pharm.* **2018**, *552*, 328–339. [CrossRef] [PubMed]
82. Konduru, N.V.; Molina, R.M.; Swami, A.; Damiani, F.; Pyrgiotakis, G.; Lin, P.; Andreozzi, P.; Donaghey, T.C.; Demokritou, P.; Krol, S.; et al. Protein Corona: Implications for Nanoparticle Interactions with Pulmonary Cells. *Part. Fibre Toxicol.* **2017**, *14*, 1–12. [CrossRef] [PubMed]
83. Hadjidemetriou, M.; Mcadam, S.; Garner, G.; Thackeray, C.; Knight, D.; Smith, D.; Al-ahmady, Z.; Mazza, M.; Rogan, J. The Human In Vivo Biomolecule Corona onto PEGylated Liposomes: A Proof-of-Concept Clinical Study. *Adv. Mater.* **2018**, *31*, 1803335. [CrossRef]
84. Del Pino, P.; Pelaz, B.; Zhang, Q.; Maffre, P.; Nienhaus, G.U.; Parak, W.J. Protein Corona Formation around Nanoparticles—From the Past to the Future. *Mater. Horiz.* **2014**, *1*, 301–313. [CrossRef]
85. Strojan, K.; Leonardi, A.; Bregar, V.B.; Kriz, I. Dispersion of Nanoparticles in Different Media Importantly Determines the Composition of Their Protein Corona. *PLoS ONE* **2017**, *12*, e0169552. [CrossRef]
86. Stark, W.J.; Stoessel, P.R.; Wohlleben, W.; Hafner, A. Industrial Applications of Nanoparticles. *Chem. Soc. Rev.* **2015**, *44*, 5793–5805. [CrossRef]
87. Hasan, A.; Morshed, M.; Memic, A.; Hassan, S.; Webster, T.J.; Marei, H.E.S. Nanoparticles in Tissue Engineering: Applications, Challenges and Prospects. *Int. J. Nanomed.* **2018**, *13*, 5637–5655. [CrossRef]
88. Danie Kingsley, J.; Ranjan, S.; Dasgupta, N.; Saha, P. Nanotechnology for Tissue Engineering: Need, Techniques and Applications. *J. Pharm. Res.* **2013**, *7*, 200–204. [CrossRef]
89. Ellis-Behnke, R.G.; Liang, Y.X.; You, S.W.; Tay, D.K.; Zhang, S.; So, K.F.; Schneider, G.E. Nano neuro knitting: Peptide nanofiber scaffold for brain repair and axon regeneration with functional return of vision. *Proc. Natl. Acad. Sci. USA* **2006**, *103*, 5054. [CrossRef]
90. De Kwaadsteniet, M.; Botes, M.; Cloete, T.E. Application of Nanotechnology in Antimicrobial Coatings in the Water Industry. *Nano* **2011**, *6*, 395–407. [CrossRef]
91. Li, Y.; Leung, P.; Yao, L.; Song, Q.W.; Newton, E. Antimicrobial Effect of Surgical Masks Coated with Nanoparticles. *J. Hosp. Infect.* **2006**, *62*, 58–63. [CrossRef]
92. Theivasanthi, T.; Alagar, M. Anti-Bacterial Studies of Silver Nanoparticles. *arXiv* **2011**, arXiv:1101.0348.
93. Kargozar, S.; Mozafari, M. Nanotechnology and Nanomedicine: Start Small, Think Big. *Mater. Today Proc.* **2018**, *5*, 15492–15500. [CrossRef]
94. Carey, J.D. Engineering the next Generation of Large-Area Displays: Prospects and Pitfalls. *Philos. Trans. R. Soc. A Math. Phys. Eng. Sci.* **2003**, *361*, 2891–2907. [CrossRef]
95. Roy, I.; Stachowiak, M.K.; Bergey, E.J. Nonviral Gene Transfection Nanoparticles: Function and Applications in the Brain. *Nanomed. Nanotechnol. Biol. Med.* **2008**, *4*, 89–97. [CrossRef]
96. Fatima, H.; Kim, K.S. Magnetic Nanoparticles for Bioseparation. *Korean J. Chem. Eng.* **2017**, *34*, 589–599. [CrossRef]
97. Adeniyi, O.M.; Azimov, U.; Burluka, A. Algae biofuel: Current status and future applications. *Renew. Sustain. Energy Rev.* **2018**, *90*, 316–335. [CrossRef]
98. Sekoai, P.T.; Ouma, C.N.M.; du Preez, S.P.; Modisha, P.; Engelbrecht, N.; Bessarabov, D.G.; Ghimire, A. Application of Nanoparticles in Biofuels: An Overview. *Fuel* **2019**, *237*, 380–397. [CrossRef]
99. Yang, G.; Wang, J. Improving Mechanisms of Biohydrogen Production from Grass Using Zero-Valent Iron Nanoparticles. *Bioresour. Technol.* **2018**, *266*, 413–420. [CrossRef]

100. Timothy, C.J.; Kogularamanan, S.; Sthepen, L. The Next Generation of Platinum Drugs: Targeted Pt (II) Agents, Nanoparticle Delivery, and Pt(IV) Prodrugs. *Chem. Rev.* **2016**, *116*, 3436–3486.
101. Díez, P.; González-Muñoz, M.; González-González, M.; Dégano, R.M.; Jara-Acevedo, R.; Sánchez-Paradinas, S.; Piñol, R.; Murillo, J.L.; Lou, G.; Palacio, F.; et al. Functional Insights into the Cellular Response Triggered by a Bile-Acid Platinum Compound Conjugated to Biocompatible Ferric Nanoparticles Using Quantitative Proteomic Approaches. *Nanoscale* **2017**, *9*, 9960–9972. [CrossRef]
102. Gao, W.; Zhang, Y.; Zhang, Q.; Zhang, L. Nanoparticle-Hydrogel: A Hybrid Biomaterial System for Localized Drug Delivery. *Ann. Biomed. Eng.* **2016**, *44*, 2049–2061. [CrossRef]
103. Dizaj, S.M.; Barzegar-jalali, M.; Zarrintan, M.H.; Adibkia, K.; Lotfipour, F. Calcium Carbonate Nanoparticles as Cancer Drug Delivery System. *Expert Opin. Drug Deliv.* **2015**, *12*, 1649–1660. [CrossRef]
104. Choi, D.G.; Venkatesan, J.; Shim, M.S. Selective Anticancer Therapy Using Pro-Oxidant Drug-Loaded Chitosan—Fucoidan Nanoparticles. *Int. J. Mol. Sci.* **2019**, *20*, 3220. [CrossRef]
105. Tatini, F.; Landini, I.; Scaletti, F.; Massai, L.; Centi, S.; Ratto, F.; Nobili, S.; Romano, G.; Fusi, F.; Messori, L.; et al. Size Dependent Biological Profiles of PEGylated Gold Nanorods. *J. Mater. Chem. B* **2014**, *2*, 6072–6080. [CrossRef]
106. De Clercq, E. Highlights in Antiviral Drug Research: Antivirals at the Horizon. *Med. Res. Rev.* **2013**, *33*, 1215–1248. [CrossRef]
107. Guidi, F.; Landini, I.; Puglia, M.; Magherini, F.; Gabbiani, C.; Cinellu, M.A.; Nobili, S.; Fiaschi, T.; Bini, L.; Mini, E.; et al. Proteomic Analysis of Ovarian Cancer Cell Responses to Cytotoxic Gold Compounds. *Metallomics* **2012**, *4*, 307–314. [CrossRef]
108. Magherini, F.; Fiaschi, T.; Valocchia, E.; Becatti, M.; Pratesi, A.; Marzo, T.; Massai, L.; Gabbiani, C.; Landini, I.; Nobili, S.; et al. Antiproliferative Effects of Two Gold(I)-N-Heterocyclic Carbene Complexes in A2780 Human Ovarian Cancer Cells: A Comparative Proteomic Study. *Oncotarget* **2018**, *9*, 28042–28068. [CrossRef]
109. Mügge, C.; Rothenburger, C.; Beyer, A.; Görls, H.; Gabbiani, C.; Casini, A.; Michelucci, E.; Landini, I.; Nobili, S.; Mini, E.; et al. Structure, Solution Chemistry, Antiproliferative Actions and Protein Binding Properties of Non-Conventional Platinum(ii) Compounds with Sulfur and Phosphorus Donors. *Dalton Trans.* **2011**, *40*, 2006–2016. [CrossRef]
110. Marzo, T.; Massai, L.; Pratesi, A.; Stefanini, M.; Cirri, D.; Magherini, F.; Becatti, M.; Landini, I.; Nobili, S.; Mini, E.; et al. Replacement of the Thiosugar of Auranofin with Iodide Enhances the Anticancer Potency in a Mouse Model of Ovarian Cancer. *ACS Med. Chem. Lett.* **2019**, *10*, 656–660. [CrossRef]
111. Maiore, L.; Cinellu, M.A.; Nobili, S.; Landini, I.; Mini, E.; Gabbiani, C.; Messori, L. Gold(III) Complexes with 2-Substituted Pyridines as Experimental Anticancer Agents: Solution Behavior, Reactions with Model Proteins, Antiproliferative Properties. *J. Inorg. Biochem.* **2012**, *108*, 123–127. [CrossRef]
112. Maiore, L.; Cinellu, M.A.; Michelucci, E.; Moneti, G.; Nobili, S.; Landini, I.; Mini, E.; Guerri, A.; Gabbiani, C.; Messori, L. Structural and Solution Chemistry, Protein Binding and Antiproliferative Profiles of Gold(I)/(III) Complexes Bearing the Saccharinato Ligand. *J. Inorg. Biochem.* **2011**, *105*, 348–355. [CrossRef]
113. Landini, I.; Lapucci, A.; Pratesi, A.; Massai, L.; Napoli, C.; Perrone, G.; Pinzani, P.; Messori, L.; Mini, E. Selection and Characterization of a Human Ovarian Cancer Cell Line Resistant to Auranofin. *Oncotarget* **2017**, *8*, 96062–96078. [CrossRef]
114. Guidi, F.; Modesti, A.; Landini, I.; Nobili, S.; Mini, E.; Bini, L.; Puglia, M.; Casini, A.; Dyson, P.J.; Gabbiani, C.; et al. The Molecular Mechanisms of Antimetastatic Ruthenium Compounds Explored through DIGE Proteomics. *J. Inorg. Biochem.* **2013**, *118*, 94–99. [CrossRef]
115. Gamberi, T.; Massai, L.; Magherini, F.; Landini, I.; Fiaschi, T.; Scaletti, F.; Gabbiani, C.; Bianchi, L.; Bini, L.; Nobili, S.; et al. ScienceDirect Proteomic Analysis of A2780/S Ovarian Cancer Cell Response to the Cytotoxic Organogold (III) Compound Aubipy C. *J. Proteom.* **2014**, *103*, 103–120. [CrossRef]
116. Sharifi, M.; Avadi, M.R.; Attar, F.; Dashtestani, F.; Ghorchian, H.; Rezayat, S.M.; Saboury, A.A.; Falahati, M. Cancer diagnosis using nanomaterials based electrochemical nanobiosensors. *Biosens. Bioelectron.* **2019**, *126*, 773–784. [CrossRef]
117. Ambrosi, A.; Airò, F.; Merkoçi, A. Enhanced Gold Nanoparticle Based ELISA for a Breast Cancer Biomarker. *Anal. Chem.* **2010**, *82*, 1151–1156. [CrossRef]
118. Ramos, M.; Castillo, C. Aplicaciones biomédicas de las nanopartículas magnéticas. *Ide@s CONCYTEG* **2011**, *6*, 629–646.

119. Haun, J.B.; Yoon, T.J.; Lee, H.; Weissleder, R. Magnetic nanoparticle biosensors. *Nanomed. Nanobiotechnol.* **2010**, *2*, 291–304. [CrossRef]
120. Riley, R.S.; June, C.H. Delivery Technologies for Cancer Immunotherapy. *Nat. Rev. Drug Discov.* **2013**, *18*, 175–196. [CrossRef]
121. Park, W.; Heo, Y.; Han, D.K. New Opportunities for Nanoparticles in Cancer Immunotherapy. *Biomater. Res.* **2018**, *22*, 24. [CrossRef]
122. Nam, J.; Son, S.; Park, K.S.; Zou, W.; Shea, L.D.; Moon, J.J. Cancer Nanomedicine for Combination Cancer Immunotherapy. *Nat. Rev. Mater.* **2019**, *4*, 398–414. [CrossRef]
123. Wang, Z.; Liu, W.; Shi, J.; Chen, N.; Fan, C. Nanoscale Delivery Systems for Cancer Immunotherapy. *Mater. Horiz.* **2018**, *5*, 344–362. [CrossRef]
124. Mi, Y.; Smith, C.C.; Yang, F.; Qi, Y.; Roche, K.C.; Serody, J.S.; Vincent, B.G.; Wang, A.Z. A Dual Immunotherapy Nanoparticle Improves T-Cell Activation and Cancer Immunotherapy. *Adv. Mater.* **2018**, *30*, 1706098. [CrossRef]
125. Surendran, S.P.; Moon, M.J.; Park, R.; Jeong, Y.Y. Bioactive Nanoparticles for Cancer Immunotherapy. *Int. J. Mol. Sci.* **2019**, *19*, 3877. [CrossRef]
126. Toy, R.; Roy, K. Engineering Nanoparticles to Overcome Barriers to Immunotherapy. *Bioeng. Transl. Med.* **2016**, *1*, 47–62. [CrossRef]
127. Mehadi, M.; Chowdhury, H.; Kubra, K.; Kanwar, R.K.; Kanwar, J.R. Nanoparticles Advancing Cancer Immunotherapy. In *Biomedical Applications of Graphene and 2D Nanomaterials*; Nurunnabi, M., McCarthy, J.R., Eds.; Elsevier: Cambridge, MA, USA, 2019; Chapter 13; pp. 283–304.
128. Cheng, C.; Castro, G.; Liu, C.; Lau, P. Advanced Nanotechnology: An Arsenal to Enhance Immunotherapy in Fi Ghting Cancer. *Clin. Chim. Acta* **2019**, *492*, 12–19. [CrossRef]
129. Hagan, C.T.; Medik, Y.B.; Wang, A.Z. Nanotechnology Approaches to Improving Cancer Immunotherapy. In *Advances in Cancer Research*; Academic Press: Cambridge, MA, USA, 2018; Volume 139, pp. 35–56.
130. Rosalia, R.A.; Cruz, L.J.; Van Duikeren, S.; Tromp, A.T.; Oostendorp, J.; Silva, A.L.; Jiskoot, W.; De Gruijl, T.; Clemens, L.; Van Der Burg, S.H.; et al. CD40-Targeted Dendritic Cell Delivery of PLGA-Nanoparticle Vaccines Induce Potent Anti-Tumor Responses. *Biomaterials* **2014**, *40*, 88–97. [CrossRef]
131. Nakamura, T.; Harashima, H. Integration of Nano Drug-Delivery System with Cancer Immunotherapy. *Ther. Deliv.* **2017**, *8*, 987–1000. [CrossRef]
132. Latterich, M.; Corbeil, J. Label-free detection of biomolecular interactions in real time with a nano-porous silicon-based detection method. *Proteome Sci.* **2008**, *6*, 31.
133. Jia, L.; Lu, Y.; Shao, J.; Liang, X.; Xu, Y. Nanoproteomics: A New Sprout from Emerging Links between Nanotechnology and Proteomics. *Trends Biotechnol.* **2013**, *31*, 99–107. [CrossRef]

nanomaterials

MDPI

Article

Highly Hydrophilic Gold Nanoparticles as Carrier for Anticancer Copper(I) Complexes: Loading and Release Studies for Biomedical Applications

Ilaria Fratoddi [1], Iole Venditti [2,*], Chiara Battocchio [2], Laura Carlini [2], Simone Amatori [1], Marina Porchia [3,*], Francesco Tisato [3], Federica Bondino [4], Elena Magnano [4], Maura Pellei [5] and Carlo Santini [5]

1 Chemistry Department Sapienza University of Rome, P.le A. Moro 5, 00185 Rome, Italy; ilaria.fratoddi@uniroma1.it (I.F.); simone.amatori@gmail.com (S.A.)
2 Sciences Department Roma Tre University of Rome, via della Vasca navale 79, 00146 Rome Italy; chiara.battocchio@uniroma3.it (C.B.); laura.carlini@uniroma3.it (L.C.)
3 ICMATE, National Research Council (CNR), Corso Stati Uniti, 4-35127 Padua, Italy; francesco.tisato@cnr.it
4 IOM-CNR Laboratorio TASC, SS 14, km 163,5 Basovizza, I-34149 Trieste, Italy; bondino@iom.cnr.it (F.B.); magnano@iom.cnr.it (E.M.)
5 School of Science and Technology, University of Camerino, 62032 Camerino (MC) Italy; maura.pellei@unicam.it (M.P.); carlo.santini@unicam.it (C.S.)
* Correspondence: iole.venditti@uniroma3.it (I.V.); marina.porchia@cnr.it (M.P.); Tel.: +39-06-5733-3388 (I.V.)

Received: 2 May 2019; Accepted: 15 May 2019; Published: 20 May 2019

Abstract: Gold nanoparticles (AuNPs), which are strongly hydrophilic and dimensionally suitable for drug delivery, were used in loading and release studies of two different copper(I)-based antitumor complexes, namely $[Cu(PTA)_4]^+$ $[BF_4]^-$ (A; PTA = 1, 3, 5-triaza-7-phosphadamantane) and $[HB(pz)_3Cu(PCN)]$ (B; $HB(pz)_3$ = tris(pyrazolyl)borate, PCN = tris(cyanoethyl)phosphane). In the homoleptic, water-soluble compound A, the metal is tetrahedrally arranged in a cationic moiety. Compound B is instead a mixed-ligand (scorpionate/phosphane), neutral complex insoluble in water. In this work, the loading procedures and the loading efficiency of A and B complexes on the AuNPs were investigated, with the aim to improve their bioavailability and to obtain a controlled release. The non-covalent interactions of A and B with the AuNPs surface were studied by means of dynamic light scattering (DLS), UV–Vis, FT-IR and high-resolution x-ray photoelectron spectroscopy (HR-XPS) measurements. As a result, the AuNPs-A system proved to be more stable and efficient than the AuNPs-B system. In fact, for AuNPs-A the drug loading reached 90%, whereas for AuNPs-B it reached 65%. For AuNPs-A conjugated systems, a release study in water solution was performed over 4 days, showing a slow release up to 10%.

Keywords: gold nanoparticles; copper(I) complexes; conjugates; drug delivery; anticancer compounds

1. Introduction

Gold nanoparticles (AuNPs) are the most versatile material in nanotechnology, with a huge range of biological and biomedical applications, such as diagnostic, therapeutic and biosensing applications [1–7]. In particular, AuNPs have been often proposed as non-toxic carriers for drug and gene-delivery applications [8–13]. In fact, the specific properties of AuNPs, such as their high surface-to-volume ratio, peculiar optical properties, easy synthesis and versatile surface functionalization, hold pledge in the clinical field for cancer therapeutics [14,15]. Moreover, AuNPs present optical properties, which can be easily tuned to desirable wavelengths according to their shape (e.g., nanoparticles, nanoshells, nanorods, etc.), size (e.g., 1 to 100 nm) and composition (e.g., core/shell or alloy noble metals) [16–20], enabling their imaging and photothermal applications [21–26]. AuNPs can also be easily functionalized with different moieties, such as antibodies, peptides and/or DNA/RNA to specifically target different cells [10,27,28], and with biocompatible molecules to prolong their in vivo circulation for drug delivery applications [29,30]. Furthermore, it is well known that passive targeting can be achieved by using AuNPs as a carrier, because of their preferential accumulation in tumor cells (enhanced permeability and retention (EPR) effect) [21].

In recent years, the biomedical research of new metal-based anticancer drugs alternative to Pt(II) derivatives (cisplatin, oxaliplatin and carboplatin, which are currently utilized in clinical practice) has been focused on complexes including, among other metals, gold, ruthenium, silver and copper [31–36]. The purpose of these studies is to circumvent severe toxicity in non-tumor cells as well as inherited and/or acquired resistance phenomena caused by Pt(II) drugs [37–39]. In particular, among the abovementioned metals, copper is receiving increasing attention [36]. Copper, as an essential micronutrient in mammalians, plays a pivotal role in redox-chemistry, growth and development, and is a key co-factor for the function of several enzymes involved in energy metabolism, respiration and DNA synthesis [40]. In addition, homeostatic mechanisms strictly define the concentration of copper in mammalian cells, which have also developed a physiological active transport mechanism for its internalization based on a trans-membrane protein referred to as human copper transporter 1 (hCtr1) [41,42]. Novel copper-based antitumor agents have been studied according to the view that endogenous metals may be less toxic toward normal cells with respect to cancer cells. Since the generally assessed mechanism of copper cell uptake implies the reduction from copper(II) to copper(I) followed by internalization through transmembrane transporters [43,44], our research has been mainly focused on copper(I) derivatives. The synthetic strategy utilizes ligands with soft donor atoms such as phosphorous in tertiary phosphanes or aromatic sp^2 hybridized nitrogen of pyrazolyl derivatives. Among these compounds, homoleptic, cationic phosphane complexes well match the ability of hCtr1 protein to internalize specifically monovalent ions, thus leading to outstanding cytotoxic efficiency toward cancer cells in both in vitro and in vivo trials [45–49]. In addition, neutral mixed-ligand complexes containing both scorpionate-like (N-donor) and phosphane ligands showed remarkable cytotoxic activity in in vitro and in vivo tests as well [50]. Despite the promising results, open problems remain, such as the low solubility and bioavailability of some of these compounds and their uncontrolled release. In this preliminary work, hydrophilic AuNPs were synthesized and loaded with either a representative of a water-soluble, cationic complex—$[Cu(PTA)_4]^+$ $[BF_4]^-$ (A; PTA = 1,3,5-triaza-7-phosphadamantane)—or a lipophilic, neutral complex—$[HB(pz)_3Cu(PCN)]$ (B; $HB(pz)_3$ = tris(pyrazolyl)borate, PCN = tris(cyanoethyl)phosphane)—aiming at the construction of a novel drug delivery system. The use of hydrophilic AuNPs as a vehicle for copper complexes is an innovative and strategic approach to improve the solubility and stability in water of the copper complexes, and consequently to increase their bioavailability. Moreover, these drug delivery systems allow the investigation of a slow and controlled release of copper complexes, opening the way for promising scenarios of in vivo and in vitro experimentation.

2. Materials and Methods

2.1. Materials and Characterizations

Sodium 3-mercapto-1-propanesulfonate ($HS(CH_2)_3SO_3Na$, 3MPS, Aldrich, 99%, St. Louis, MO, USA), tetrachloroauric(III) acid trihydrate ($HAuCl_4{\cdot}3H_2O$, Aldrich, 99.0%, St. Louis, MO, USA), sodium borohydride ($NaBH_4$) and PBS buffer solution at pH = 7.4 (technical grade Aldrich, St. Louis, MO, USA) were used as received. UV–Vis spectra were acquired in H_2O and MeOH solutions by using quartz cells with a Varian Cary 100 Scan UV–Vis spectrophotometer. The size distribution of AuNPs in H_2O solution was investigated by means of the dynamic light scattering (DLS) technique by using a Zetasizer Nanoseries Malvern instrument, at the specific temperature (25.0 ± 0.2 °C and 37.0 ± 0.2 °C). Correlation data were acquired and fitted in reference to our previous work [51,52]. Field emission scanning electron microscopy (FESEM) images were acquired with an Auriga Zeiss instrument, resolution 1 nm, applied voltage 6–12 kV) on freshly prepared films drop-cast from a water solution on a metallic sample holder. A Mini Spin-Eppendorf centrifuge was used for the purification of AuNPs samples (13,000 rpm, 20 min, five times with deionized water). Deionized water was obtained from Zeener Power I Scholar-UV (18.2 MΩ). A Scanvac-CoolSafe55-4 Lyophilizer was used to dry samples. Attenuated total reflection (ATR) spectra were recorded with a Bruker Vertex 70 instrument in the range of 4000–400 cm^{-1}. High-resolution x-ray photoelectron spectroscopy (HR-XPS) experiments were carried out at the CNR BACH (Beam line for Advanced DiCHroism) [53] undulator beam line at the ELETTRA Synchrotron Radiation facility of Trieste (Italy). XPS data were collected with a pass energy equal to 50 eV, with the monochromator entrance and exit slits fixed at 20 μm. Photon energies of 386 eV, 596 eV and 1022 eV were used respectively for C1s, Au4f, P2p, B1s; O1s, N1s; Cu2p spectral regions, with an energy resolution of 0.23 eV. C1s spectra used for calibration were recorded at all photon energies. Calibration of the energy scale was made by referencing all the spectra to the C1s core level signal of aliphatic C atoms, always found at 285.00 eV, and the Au4f7/2 signal of metallic gold measured on a reference gold foil (84.00 eV Binding Energy, BE) [54,55]. A curve-fitting analysis was performed using Gaussian curves as fitting functions. When several different species were individuated in a spectrum, the same FWHM value was used for all individual photoemission bands. To perform HR-XPS analysis, pristine Cu(I) complexes (A = $[Cu(PTA)_4]^+$ $[BF_4]^-$ and B = $[HB(pz)_3Cu(PCN)]$ and functionalized Au nanoparticles (AuNPs-A and AuNPs-B) were deposited onto TiO_2/Si(111) (as to avoid any signal interference) wafer substrates by following a drop-casting procedure.

2.2. Preparation of Conjugate Nanoparticles

The AuNPs stabilized with 3MPS were synthesized as previously reported [9]. Briefly, starting with 200 mg (5×10^{-4} mol) of $HAuCl_4 \times 3H_2O$ in 20 mL of deionized water, a solution of 3MPS in 20 mL of deionized water was added under vigorous stirring (Au/S = 1/4 molar ratios). Two hours after the addition of the reduction agent (a water solution of $NaBH_4$, Au/$NaBH_4$ molar ratio = 1/10), a solid black product was isolated and purified by centrifugation (13,000 rpm, 20 min, five times with deionized water). AuNP main characterizations: UV–Vis (λ_{max} (nm), H_2O) 523 nm; DLS ($<2R_H>$ (nm), H_2O): 12 ± 3 nm; Z potential: −35 ± 3 mV; FESEM (nm) 8–10 nm. Complexes A and B were prepared according to procedures reported in the literature [45,50]. They showed a well-known UV–Vis spectrum, as reported in the Supporting Information, Figures S1a,b). The conjugate nanoparticles were prepared using the following procedure. The AuNPs and A were mixed in water (Au/A = 5/1 w/w) under gentle stirring (room temperature, 4 h) and then the suspension was centrifuged (13,000 rpm, 2 h) to obtain AuNPs-A as a solid residue. It was lyophilized and stored at room temperature, while the supernatant was used for loading evaluations. The AuNPs and B were mixed in MeOH (Au/B= 5/1 w/w) under gentle stirring (room temperature, 4 h) and then the suspension was centrifuged (13,000 rpm, 30 min) to obtain AuNPs-B as a solid residue. It was lyophilized and stored at room temperature, while the supernatant was used for loading evaluations. The loading and the loading efficiency (η) were calculated by using calibration curves, as reported in the Supporting Information (Figures S1c,d) and calculated in reference to our previous work [9]. For each sample, three independent measurements were carried out and the mean value and standard deviation are reported.

2.3. Stability and Release Studies

For the stability studies, AuNPs-A and AuNPs-B were dispersed in water at the concentration of 0.1 mg/mL and the size of nanoparticles was measured over 10 days at room temperature. The release studies were performed in H_2O at 37 °C over 4 days, using 2 mg of conjugated nanoparticles in 20 mL of media. The released was calculated by analyzing the water solution and detecting the free complex by using UV–Vis measurements, in reference to our previous work [9]. For each sample, three independent measurements were carried out and the mean value and standard deviation are reported.

3. Results and Discussion

3.1. Conjugate Nanoparticles: Preparation, Characterization and Loading Studies

Highly hydrophilic gold nanoparticles were synthetized following a previously published procedure [9], and the UV–Vis spectrum and DLS measurements confirmed their nanodimension, as shown in Figure 1a,b. The results revealed that these AuNPs are particularly suitable for Cu(I) complexes delivery. In fact, AuNPs functionalized by 3MPS showed a high degree of stability and hydrophilicity due to the small length alkyl chains thiol with a charged terminal sulphonate group. Moreover, the 3MPS choice as a ligand, with a molar ratio of Au/S = $\frac{1}{4}$, guarantees a balance between stability and loading and favours transport in a watery environment. This fact increases the final bioavailability of the conjugates, especially for compounds with low water solubility. Furthermore, the plasmonic absorption peaks of the AuNPs and the absorption peaks of the copper complexes (λ_{max} at 228 nm and 268 nm for complexes A and B, respectively, as shown in Figure 1a and in Figures S1a,b [45,50]) appeared in different areas of the spectrum—in the UV spectrum for complexes and in the visible spectrum for AuNPs, allowing easy detection of the loading processes. This feature makes it possible to design a loading protocol based on the simple physical contact of AuNPs and copper complexes that can be physically adsorbed.

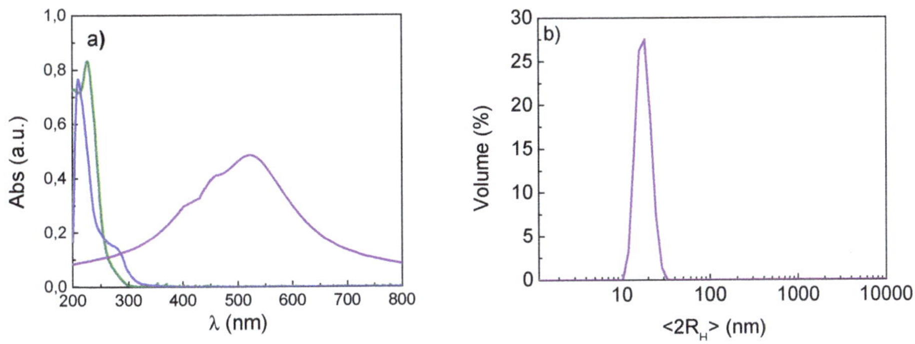

(a) (b)

Figure 1. (**a**) UV–Vis spectrum of gold nanoparticles (AuNPs) (violet curve) and complexes A (green curve) and B (blue curve); (**b**) dynamic light scattering (DLS) measurement in water of AuNPs alone (in violet): $<2R_H> = 15 \pm 2$ nm.

On the basis of these considerations, the loading protocol for AuNPs and the two Cu(I) complexes was performed in a water solution at room temperature under gentle stirring. In Figure 2a,b, the chemical structures of the anticancer Cu(I) complexes used in this study, A and B, and a sketch of the loading protocol to obtain AuNPs-A and AuNPs-B conjugates are reported. The value of the loading efficiency η (%), reported in Figure 2c, can be calculated as follows [9,13]:

$$\eta\ (\%) = (m_{loaded\ drug}/m_{drug})\ 100,$$

where $m_{loaded\ drug}$ is the mass of the loaded drug (A or B), calculated from UV–Vis quantitative data and m_{drug} is the mass of the drug (A or B) used in the experimental procedure. From the absorbance value of free A or B, it is possible to obtain the amount of loaded drug, by determining the difference.

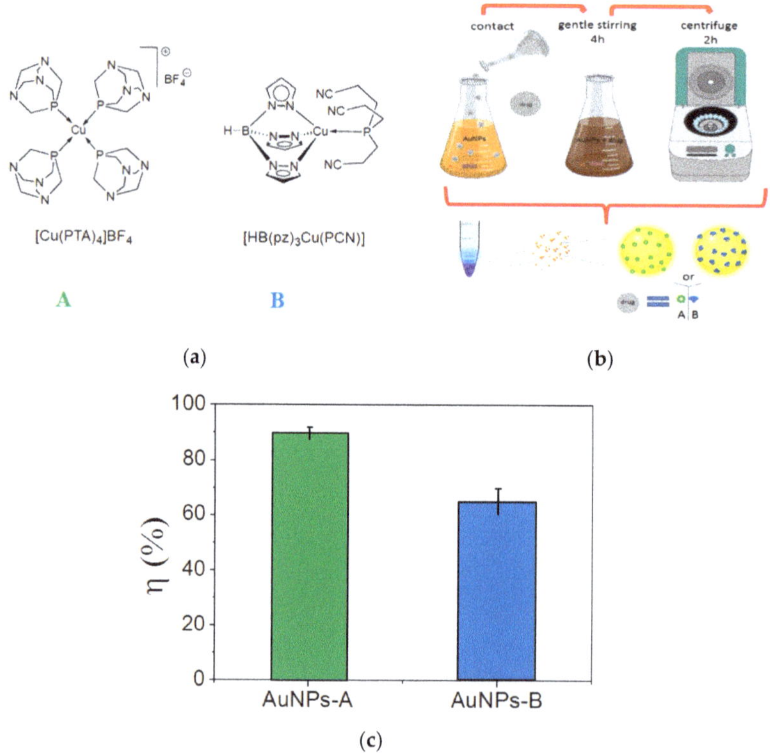

Figure 2. (**a**) Chemical structures of anticancer Cu(I) complexes used in this study; (**b**) sketch of loading protocol to obtain AuNPs-A and AuNPs-B conjugates; (**c**) loading efficiency η (%) for AuNPs-A (in green, η (%) = 90 ± 4 %) and AuNPs-B (in blue, η (%) = 65 ± 10 %).

The loading studies allowed to us obtain the conjugated systems: AuNPs-A with η = 90 ± 4% and AuNPs-B with η (%) = 65 ± 10%. These systems were characterized in depth to understand the chemical interaction between AuNPs and Cu(I) complexes A and B. DLS studies were performed in a water suspension and showed a dimensional increase for AuNPs-A and AuNPs-B compared with the AuNPs alone, as reported in Figure 3a. In fact, the conjugation of the complexes involved a different degree of hydration of the particles and, as a result, the hydrodynamic diameter (<$2R_H$>) increased.

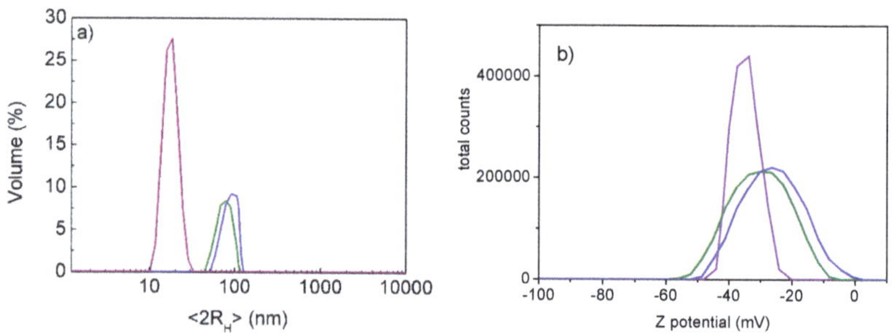

Figure 3. DLS data of AuNPs in violet, AuNPs-A in green and AuNPs-B in blue: (**a**) <$2R_H$> in water: AuNPs (15 ± 2 nm), AuNPs-A (56 ± 30 nm) and AuNPs-B (76 ± 32 nm); (**b**) Z potential in water: AuNPs (−35 ± 2 mV), AuNPs-A (−30 ± 3 mV) and AuNPs-B (−23 ± 4 mV).

Moreover, DLS allowed us to measure the electrophoretic mobility and, using the Smoluchowski equation, the Z potential [51,52]. The Z potential is the potential at this slipping plane, i.e., the surrounding electrical double layer, where the liquid moves together with particles. Therefore, the measured Z potential is not exactly the surface potential (surface charge density), but it is the potential of practical interest because it determines the inter-particle forces and enables the evaluation of the stability of the colloidal system [51,52]. The Z potential studies performed on our conjugates systems confirmed these interactions between AuNPs and complexes A and B. In fact, the Z potential was -30 ± 3 mV and -23 ± 4 mV, respectively, for AuNPs-A and AuNPs-B, instead of -35 ± 2 mV for AuNPs alone, as shown in Figure 3b. This is due to two different effects introduced by the presence of complexes A and B on the surface of the AuNPs. The first effect is the decrease of the negative charge density, due to the presence of neutral or positively charged molecules on the gold surface; this is strictly related to the Stern layer and slipping plane around the nanoparticles, which produce the Z potential. The second effect is the different aggregation grade, also observable from signal enlargement and size measurements, due to the interaction between complex molecules linked on different and vicinal AuNPs, which cause the system to be less stable in general. The balance or the prevalence of one of these two effects also explains the slight difference between the Z potential values of AuNPs-A and AuNPs-B. In fact, the conjugate AuNPs-B showed a larger size and lower Z potential with respect to AuNPs-A (see Figure 3). Indeed, complex A had a positive charge, facilitating adsorption and producing greater loading efficiency. Moreover, complex A on AuNPs decreased the interaction phenomena between the absorbed complex molecules, reducing the aggregation phenomena of the colloidal system in a solution. This justifies the DLS results regarding the smaller dimensions and more negative Z potential of AuNPs-A compared to those of AuNPs-B.

FESEM-EDX investigations performed on conjugate AuNPs-A nanoparticles showed dimensions around 10 nm with the presence of some aggregates (see Figure S2). It can be noticed that the dimensions obtained from DLS were greater than those obtained from FESEM images. Such a dimensional difference is due to the intrinsic difference between the two techniques, based on different principles. In fact, DLS estimated the particles hydrodynamic diameter ($<2R_H>$) in the aqueous environment, with the important effect of swelling, and this dimension is the Z average value, which is the mean diameter weighted over the scattered light intensity. On the other hand, microscopy measurements were carried out under a vacuum on a dry sample deposited by casting with no hydration effects. The particles were more or less aggregated due to concentration or to fast or slow solvent evaporation occurring during the preparation of the sample. For this reason, it is difficult to directly compare the two measurements.

The Energy Dispersive X-ray Analysis (EDX) evidenced the presence of Cu and, in particular, the semiquantitative analysis showed the ratio of Au:Cu to be around 0.4:0.03 (see Figure S2).

The ATR data confirmed the effective interaction between copper complexes and AuNPs. In fact, both for AuNPs-A and AuNPs-B, typical bands were found (see Figures S3a,b). Particularly for AuNPs-A, some characteristic signals of the A complex were recognizable, such as the bending of CH_2 at 1456 and 1418 cm^{-1} and the C–N stretching at 1043 and 1000 cm^{-1}, thus confirming the successful conjugation. A shift of the C–N stretching signals was also observed, which moved from 1103 and 947 cm^{-1} in the free complex to 1043 and 1000 cm^{-1} in the conjugate, suggesting a direct involvement of these groups in the interaction with the gold nanoparticle surface. For AuNPs-B, the ATR measurements showed the typical bands at 2480 cm^{-1} due to the B–H stretching, bands at 1502 and 1403 cm^{-1} due to the C–C bond of pyrazole rings and at 1301 cm^{-1} due to C–N stretching. In this case the nitrile stretching showed a shift from 2254 cm^{-1} (free complex) to 2240 cm^{-1} (conjugate system), highlighting the involvement of these groups in the conjugation formation [33].

3.2. HR-XPS Studies

Molecular and electronic structure of AuNPs-Cu(I) complexes assemblies were probed by means of synchrotron radiation-induced photoemission spectroscopy (HR-XPS); for comparison, the pristine Cu(I) complexes A and B ($[Cu(PTA)_4]^+$ $[BF_4]^-$ and $[HB(pz)_3Cu(PCN)]$, respectively) were also investigated. Signals were acquired at C1s, P2p, N1s, B1s, Cu2p and Au4f core levels, and the obtained spectra were analyzed by following a peak fitting procedure that evidenced spectral components arising from atoms in different electronic environments.

C1s and P2p spectra data analysis allowed us to assess the Cu(I) complex stability; as reported in Table S1 in the Supporting Information, P2p and C1s components positions, i.e., BE values, which reflect the molecular composition of the organic ligands, were not affected by the AuNP-Cu(I) complexes interaction. Indeed, the $P2p_{3/2}$ spin-orbit component of the phosphorous signal was always found around 131.50 eV BE (complex A: 131.03 eV; AuNP-A: 131.11 eV; complex B: 131.81 eV; AuNP-B 131.86 eV), as expected from the literature for organic molecules containing P atoms [56]. The observed P2p BE stability allowed us to completely dismiss the occurrence of any degradation effect due to molecule oxidation, which would result in phosphane oxide formation with a noticeable shift in the P2p signal BE towards higher values [56]). P2p spectra are reported in Figure S4 in the Supporting Information. Au4f spectra appeared composed, showing a spin-orbit peak of high intensity due to metallic gold atoms at the nanoparticle cores ($Au4f_{7/2}$ BE = 83.9 eV), and a second signal of low intensity at higher BE values ($Au4f_{7/2}$ BE = 85.1 eV) was associated with partially positively charged gold atoms at the NP surface, as expected from the literature on analogous systems [4]. Copper Cu2p core signals were also acquired; for all samples, a single spin-orbit pair was observed, compatible with Cu(I) ions ($Cu2p_{3/2}$ = 931.5 eV BE) [56]. The Au4f spectrum of AuNP-A and the Cu2p spectra of both AuNP-A and complex A were reported as examples in the Supporting Information (Figure S5).

The most interesting signal that shed light into the Cu(I) complex/AuNP interaction was the N1s core level. As reported in Table 1, both Cu(I) compounds showed N1s signals at the BE, as expected in the literature for the proposed molecular structure (tertiary amines N1s are expected at about 400 eV BE, and was found at 399.99 eV for complex A; $N\equiv C-R$ like nitrogen N1s signal is expected at 399.6 eV BE [57], and was found at 399.7 eV BE in complex B [56]). In Figure 4 all N1s spectra are collected. Contributions related to pristine nitrogen atoms in Cu(I) complexes are represented in red.

Table 1. N1s BE, FWHM values and assignments for pristine Cu(I) complexes and AuNPs carriers.

Sample	BE (eV)	FWHM (eV)	Assignment
A	399.99	1.64	NR_3
AuNPs-A	399.72	1.44	NR_3
	400.95	1.44	NR_3H^+
B	399.71	2.00	$N\equiv C-R$
AuNPs-B	398.70	2.06	$N\equiv C-R(\delta^-)$
	399.70	2.06	$N\equiv C-R$

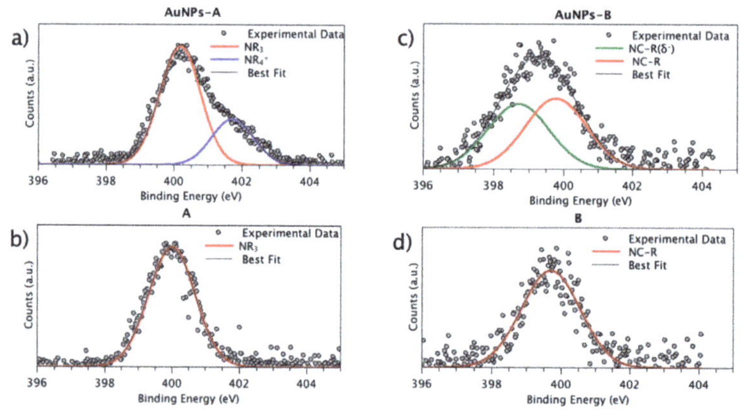

Figure 4. High-resolution x-ray photoelectron spectroscopy (HR-XPS) N1s spectra of: (**a**) AuNPs-A; (**b**) complex A; (**c**) AuNPs-B; (**d**) complex B.

As evidenced in Table 1 and clearly observable in Figure 4, when Cu(I) complexes (Figure 4b,d) interacted with the gold nanoparticles, the N1s signal shape was modified (Figure 4a,c). For AuNP-A, a shoulder appeared at high BE (Figure 4a), suggesting a new spectral component at about 401 eV BE (in blue in the figure), which is usually assigned to positively charged N atoms in quaternary ammonium salts. On the other hand, the N1s spectrum of AuNP-B was larger at low BE, suggesting that a second spectral component appeared at a lower BE than the pristine $N \equiv C - R$-like N atom (Figure 4c); this behaviour indicates a partial electron transfer from the AuNPs to the nitrogen atom of the $N \equiv C - R$ functional group.

3.3. Conjugate Nanoparticles: Stability and Release Studies

It must also be highlighted that the main advantage of drug delivery with AuNPs is the possibility of studying a targeted and controlled release in terms of target site and time, as reported in many recent papers [8–10]. In this context, it is important to verify the stability of conjugates, also in view of a future formulation that makes them easily storable and at the same time quickly ready for use. The lyophilization appears to fulfil to this requirement, and was therefore performed by studying the effects on AuNPs alone and conjugates, both in terms of dimensional and structural stability. Comparing the fresh samples and the lyophilized samples by DLS measurements, aggregation phenomena were observed with important dimensional variations, as shown in Figure 5a–c. These phenomena involved both AuNPs alone and conjugated systems that remained however with dimensions under 300 nm. Regarding the structural stability of the systems, they showed unaltered structural conformation, confirmed by FTIR measurements, and good results in terms of $<2R_H>$ reproducibility when they were re-suspended in water at room temperature up to 10 days, as reported in Figure 5d.

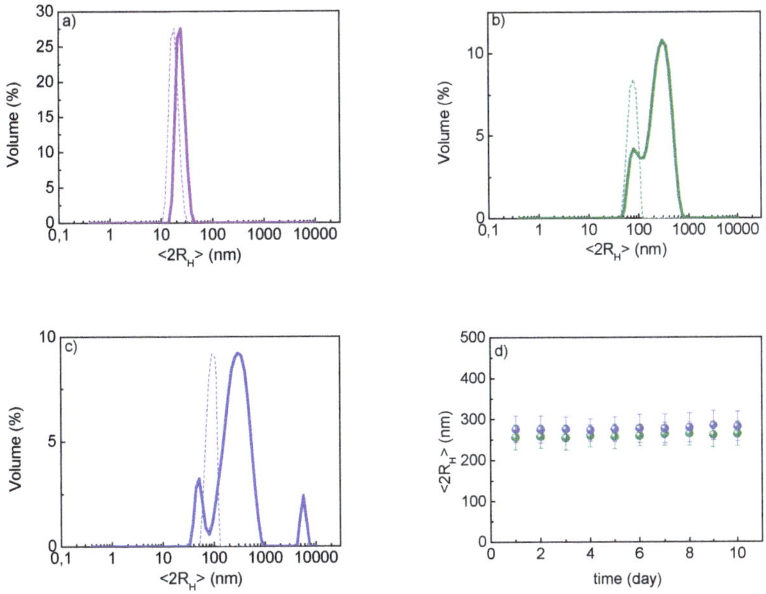

Figure 5. DLS stability study in water of AuNPs in violet, AuNPs-A in green and AuNPs-B in blue: (**a**) AuNPs $<2R_H>$ before (dashed line) 15 ± 2 nm and after samples lyophilization (solid line) 25 ± 5 nm; (**b**) AuNPs-A $<2R_H>$ before (dashed line) 56 ± 30 nm and after lyophilization (solid line) 256 ± 30 nm; (**c**) AuNPs-B $<2R_H>$ before (dashed line) 76 ± 32 nm and after lyophilization (solid line) 276 ± 32 nm; (**d**) $<2R_H>$ of lyophilized AuNPs-A and AuNPs-B re-suspended in water at different days, up to 10 days.

Therefore, the study on the release was performed. The conjugate was re-suspended in water at 37 °C with gentle stirring and the UV–Vis analysis of the aqueous solution at defined times allowed the quantification of the released complex. This study showed strong and different interactions between Cu(I) complexes A and B and AuNPs, as already evidenced by the HR-XPS studies. In particular, for AuNPs-A, the interaction involved nitrogen that partially transferred electrons to the surface of the metal, creating an interaction that caused a slow release. In Figure 6 the slow release profile of AuNPs-A, less than 10% up to 4 days, is shown. On the other hand, the AuNPs-B system involved the N ≡ C − R moiety, which strongly interacted with the gold surface, making the release not appreciable in a few days (up to 4). Taking these results into account, surely the AuNPs-A conjugate is more promising, not only for the better loading efficiency but, above all, for the evident slow release, unlike the AuNPs-B conjugate.

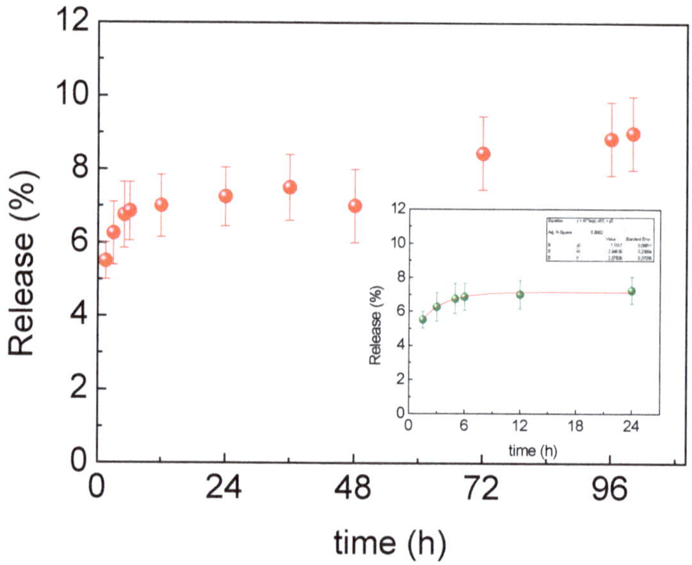

Figure 6. Released profile of AuNPs-A with inset details showing the release in time in the range of 0–24 h.

AuNPs-A showed an excellent and promising result, because with a single administration it could be possible to achieve a slow drug delivery release in a biological site up over 4 days and more. Moreover, a main advantage of delivering a water-soluble drug with AuNPs is the accumulation of AuNPs in cancer cells, which guarantees the drug's targeting. Further, the slow release is an excellent opportunity to study the synergistic effects of AuNPs and copper complexes, effects that could occur for a long time (days and weeks), as in the case of slow-release anti-inflammatory drugs reported in the literature [9]. This fact opens new scenarios for investigations related to the action mechanisms as well as for synergistic action with AuNPs-A.

4. Conclusions

Strongly hydrophilic gold nanoparticles, AuNPs, were prepared to be conjugated with copper(I) complexes. In particular, loading and release studies were performed using two different copper(I) antitumor complexes, namely $[Cu(PTA)_4]^+$ $[BF_4]^-$ (A; PTA = 1,3,5-triaza-7-phosphadamantane) and $[HB(pz)_3Cu(PCN)]$ (B; $HB(pz)_3$ = tris(pyrazolyl)borate, PCN = tris(cyanoethyl)phosphane). In the water-soluble compound A, the metal is tetrahedrally arranged in a cationic moiety, while compound B is a mixed-ligand (scorpionate/phosphane), neutral complex insoluble in water. Loading protocols and efficiency are also related to these structural aspects and were optimized to obtain $\eta = 90 \pm 4\%$ and $\eta = 65 \pm 10\%$, respectively, for AuNPs-A and AuNPs-B. Structural differences of A and B induced different behaviours regarding the interactions with the gold surface, as showed by the HR-XPS studies. In fact, for compound A, nitrogen partially transfers electrons to the surface of the metal nanoparticles, creating an interaction that causes a slow release in water, less than 10% in 4 days. On the other hand, in B compound the $N \equiv C - R$ groups hook onto the surface of the gold, producing a strong interaction that makes the release not appreciable in the same time interval (up to 4 days). Therefore, both AuNPs-A and AuNPs-B represent promising examples of water-soluble gold nanocarriers suitable to improve the bioavailability of synthetic drugs, especially considering the EPR effect of AuNPs. In particular AuNPs-A, which achieved a slow release, opens the way for biological in vitro studies to explore the synergic activity of copper complexes and gold nanoparticles.

Supplementary Materials: The following are available online at http://www.mdpi.com/2079-4991/9/5/772/s1, Figure S1: Uv–Vis spectra and calibration curves for complex A (green) and complex B (blue); Figure S2: SEM-EDX analysis on AuNPs-A; Figure S3: ATR data of conjugates systems: (a) AuNPs-A (PTA); (b) AuNPs-B (PCN); Table S1. C1s and P2p spectra data analysis BE, FWHM values and assignments for pristine Cu(I) complexes and AuNPs carriers; Figure S4: XPS P2p spectra confirming the molecular structure stability of A and B complexes; Figure S5: (a) XPS Au4f spectrum of AuNP-A; (b) Cu2p spectra of complex A and AuNP-A (rough data, confirming the stability of Cu(I) complex).

Author Contributions: I.F. and I.V. designed and made the chemical experimental synthesis with AuNPs and conjugate systems, and performed the DLS measurements and evaluations with S.A. M.P. (Marina Porchia), F.T. and M.P. (Maura Pellei), C.S. performed synthesis and characterizations of Cu(I) complexes and the relative data analysis. L.C. and C.B. performed the H-XPS measurements and the relative data analysis. S.N. and E.M. provided technical support in SR-XPS measurements. All authors contributed to the paper writing.

Funding: This research received no external funding.

Acknowledgments: Sapienza authors acknowledged Ateneo Sapienza 2017 project. XPS measurements at the ELETTRA facility were supported within project N. 20160181. The Grant of Excellence Departments, MIUR (ARTICOLO 1, COMMI 314–337 LEGGE 232/2016) is gratefully acknowledged by the authors of Roma Tre University. We are grateful to CIRCMSB (Consorzio Interuniversitario di Ricerca in Chimica dei Metalli nei Sistemi Biologici). Images in Figures 2b and 3 were generated with BioRender.com

Conflicts of Interest: The authors declare no conflict of interest.

References

1. McQuaid, H.N.; Muir, M.F.; Taggart, L.E.; McMahon, S.J.; Coulter, J.A.; Hyland, W.B.; Jain, S.; Butterworth, K.T.; Schettino, G.; Prise, K.M.; et al. Imaging and radiation effects of gold nanoparticles in tumour cells. *Sci. Rep.* **2016**, *6*, 19442. [CrossRef]
2. Zhou, W.; Gao, X.; Liu, D.; Chen, X. Gold Nanoparticles for In Vitro Diagnostics. *Chem. Rev.* **2015**, *115*, 10575–10636. [CrossRef] [PubMed]
3. Fu, Q.; Wu, Z.; Xu, F.; Li, X.; Yao, C.; Xu, M.; Sheng, L.; Yu, S.; Tang, Y. A portable smart phone-based plasmonic nanosensor readout platform that measures transmitted light intensities of nanosubstrates using an ambient light sensor. *Lab Chip* **2016**, *16*, 1927–1933. [CrossRef] [PubMed]
4. Venditti, I.; Hassanein, T.F.; Fratoddi, I.; Fontana, L.; Battocchio, C.; Rinaldi, F.; Carafa, M.; Marianecci, C.; Diociaiuti, M.; Agostinelli, E.; et al. Bioconjugation of gold-polymer core–shell nanoparticles with bovine serum amine oxidase for biomedical applications. *Colloids Surf. B* **2015**, *134*, 314–321. [CrossRef]
5. Xie, J.; Lee, S.; Chen, X. Nanoparticle-based theranostic agents. *Adv. Drug Deliv. Rev.* **2010**, *62*, 1064–1079. [CrossRef] [PubMed]
6. Yu, M.; Lei, B.; Gao, C.; Yan, J.; Ma, P.X. Optimizing surface-engineered ultra-small gold nanoparticles for highly efficient miRNA delivery to enhance osteogenic differentiation of bone mesenchymal stromal cells. *Nano Res.* **2017**, *10*, 49–63. [CrossRef]
7. Beik, J.; Abed, Z.; Ghadimi-Daresajini, A.; Nourbakhsh, M.; Shakeri-Zadeh, A.; Ghasemi, M.S.; Shiran, M.B. Measurements of nanoparticle-enhanced heating from 1MHz ultrasound in solution and in mice bearing CT26 colon tumors. *J. Therm. Biol.* **2016**, *62*, 84–89. [CrossRef]

8. Yavuz, M.S.; Cheng, Y.; Chen, J.; Cobley, C.M.; Zhang, Q.; Rycenga, M.; Xie, J.; Kim, C.; Song, K.H.; Schwartz, A.G.; et al. Gold nanocages covered by smart polymers for controlled release with near-infrared light. *Nat. Mater.* **2009**, *8*, 935–939. [CrossRef]
9. Rossi, A.; Donati, S.; Fontana, L.; Porcaro, F.; Battocchio, C.; Proietti, E.; Venditti, I.; Bracci, L.; Fratoddi, I. Negatively charged gold nanoparticles as a dexamethasone carrier: Stability in biological media and bioactivity assessment: In vitro. *RSC Adv.* **2016**, *6*, 99016–99022. [CrossRef]
10. Fitzgerald, K.A.; Rahme, K.; Guo, J.; Holmes, J.D.; O'Driscoll, C.M. Anisamide-targeted gold nanoparticles for siRNA delivery in prostate cancer—synthesis, physicochemical characterisation and in vitro evaluation. *J. Mater. Chem. B* **2016**, *4*, 2242–2252. [CrossRef]
11. Fratoddi, I.; Venditti, I.; Cametti, C.; Russo, M.V. The puzzle of toxicity of gold nanoparticles. The case-study of HeLa cells. *Toxicol. Res.* **2015**, *4*, 796–800. [CrossRef]
12. Nishiyama, N. Nanocarriers shape up for long life. *Nat. Nanotechnol.* **2007**, *2*, 203–204. [CrossRef]
13. Fratoddi, I.; Benassi, L.; Vaschieri, C.; Venditti, I.; Bessar, H.; Mai, S.; Azzoni, P.; Magnoni, C.; Costanzo, A.; Casagrande, V.; et al. Effects of topical methotrexate loaded gold nanoparticle in cutaneous inflammatory mouse model. *Nanomed. NBM Nanotechnol. Biol. Med.* **2019**, *17*, 276–286. [CrossRef]
14. Tummala, S.; Kumar, M.N.S.; Pindiprolu, S.K. Improved anti-tumor activity of oxaliplatin by encapsulating in anti-DR5 targeted gold nanoparticles. *Drug Deliv.* **2016**, *23*, 3505–3519. [CrossRef]
15. Her, S.; Jaffray, D.A.; Allen, C. Gold nanoparticles for applications in cancer radiotherapy: Mechanisms and recent advancements. *Adv. Drug Deliv. Rev.* **2017**, *109*, 84–101. [CrossRef]
16. Zhao, F.; Li, X.; Li, J.; Dou, Y.; Wang, L.; Wu, M.; Liu, Y.; Chang, J.; Zhang, X. Activatable ultrasmall gold nanorods for "off–on" fluorescence imaging-guided photothermal therapy. *J. Mater. Chem. B* **2017**, *5*, 2145–2151. [CrossRef]
17. Venditti, I. Gold nanoparticles in photonic crystals applications: A review. *Materials* **2017**, *10*, 97. [CrossRef]
18. Cai, K.; Zhang, W.; Zhang, J.; Li, H.; Han, H.; Zhai, T. Design of Gold Hollow Nanorods with Controllable Aspect Ratio for Multimodal Imaging and Combined Chemo-Photothermal Therapy in the Second Near-Infrared Window. *ACS Appl. Mater. Interfaces* **2018**, *10*, 36703–36710. [CrossRef]
19. Karabel Ocal, S.; Patarroyo, J.; Kiremitler, N.B.; Pekdemir, S.; Puntes, V.F.; Onses, M.S. Plasmonic assemblies of gold nanorods on nanoscale patterns of poly(ethylene glycol): Application in surface-enhanced Raman spectroscopy. *J. Colloid Interface Sci.* **2018**, *532*, 449–455. [CrossRef]
20. Placido, T.; Tognaccini, L.; Howes, B.D.; Montrone, A.; Laquintana, V.; Comparelli, R.; Curri, M.L.; Smulevich, G.; Agostiano, A. Surface Engineering of Gold Nanorods for Cytochrome. *ACS Omega* **2018**, *3*, 4959–4967. [CrossRef]
21. Mieszawska, A.J.; Mulder, W.J.M.; Fayad, Z.A.; Cormode, D.P. Multifunctional Gold Nanoparticles for Diagnosis and Therapy of Disease. *Mol. Pharm.* **2013**, *10*, 831–847. [CrossRef] [PubMed]
22. Zhang, J.; Li, J.; Kawazoe, N.; Chen, G. Composite scaffolds of gelatin and gold nanoparticles with tunable size and shape for photothermal cancer therapy. *J. Mater. Chem. B* **2017**, *5*, 245–253. [CrossRef]
23. Chen, H.; Shao, L.; Ming, T.; Sun, Z.; Zhao, C.; Yang, B.; Wang, J. Understanding the Photothermal Conversion Efficiency of Gold Nanocrystals. *Small* **2010**, *6*, 2272–2280. [CrossRef] [PubMed]
24. Chauhan, G.; Chopra, V.; Tyagi, A.; Rath, G.; Sharma, R.K.; Goyal, A.K. "Gold nanoparticles composite-folic acid conjugated graphene oxide nanohybrids" for targeted chemo-thermal cancer ablation: In vitro screening and in vivo studies. *Eur. J. Pharm. Sci.* **2017**, *96*, 351–361. [CrossRef]
25. D'Acunto, M. Detection of Intracellular Gold Nanoparticles: An Overview. *Materials* **2018**, *11*, 882. [CrossRef]
26. Zhou, B.; Song, J.; Wang, M.; Wang, X.; Wang, J.; Howard, E.W.; Zhou, F.; Qu, J.; Chen, W.R. BSA-bioinspired gold nanorods loaded with immunoadjuvant for the treatment of melanoma by combined photothermal therapy and immunotherapy. *Nanoscale* **2018**, *10*, 21640–21647. [CrossRef]
27. Yi, Y.; Kim, H.J.; Mi, P.; Zheng, M.; Takemoto, H.; Toh, K.; Kim, B.S.; Hayashi, K.; Naito, M.; Matsumoto, Y.; et al. Targeted systemic delivery of siRNA to cervical cancer model using cyclic RGD-installed unimer polyion complex-assembled gold nanoparticles. *J. Control. Release* **2016**, *244*, 247–256. [CrossRef]
28. Li, W.; Zhao, X.; Du, B.; Li, X.; Liu, S.; Yang, X.-Y.; Ding, H.; Yang, W.; Pan, F.; Wu, X.; et al. Gold Nanoparticle–Mediated Targeted Delivery of Recombinant Human Endostatin Normalizes Tumour Vasculature and Improves Cancer Therapy. *Sci. Rep.* **2016**, *6*, 30619. [CrossRef]

29. Wang, J.; Bai, R.; Yang, R.; Liu, J.; Tang, J.; Liu, Y.; Li, J.; Chai, Z.; Chen, C. Size- and surface chemistry-dependent pharmacokinetics and tumor accumulation of engineered gold nanoparticles after intravenous administration. *Metallomics* **2015**, *7*, 516–524. [CrossRef]
30. Xia, Q.; Li, H.; Xiao, K. Factors Affecting the Pharmacokinetics, Biodistribution and Toxicity of Gold Nanoparticles in Drug Delivery. *Curr. Drug Metab.* **2016**, *17*, 849–861. [CrossRef]
31. Dyson, P.J.; Sava, G. Metal-based antitumour drugs in the post genomic era. *Dalton Trans.* **2006**, *16*, 1929–1933. [CrossRef]
32. Sava, G.; Bergamo, A.; Dyson, P.J. Metal-based antitumour drugs in the post-genomic era: What comes next? *Dalton Trans.* **2011**, *40*, 9069–9075. [CrossRef]
33. Frezza, M.; Hindo, S.; Chen, D.; Davenport, A.; Schmitt, S.; Tomco, D.; Ping Dou, Q. Novel Metals and Metal Complexes as Platforms for Cancer Therapy. *Curr. Pharm. Des.* **2010**, *16*, 1813–1825. [CrossRef]
34. Monneret, C. Platinum anticancer drugs. From serendipity to rational design. *Ann. Pharm. Franc.* **2011**, *69*, 286–295. [CrossRef]
35. Banti, C.N.; Hadjikakou, S.K. Anti-proliferative and anti-tumor activity of silver(i) compounds. *Metallomics* **2013**, *5*, 569–596. [CrossRef]
36. Santini, C.; Pellei, M.; Gandin, V.; Porchia, M.; Tisato, F.; Marzano, C. Advances in Copper Complexes as Anticancer Agents. *Chem. Rev.* **2014**, *114*, 815–862. [CrossRef]
37. Rabik, C.A.; Dolan, M.E. Molecular mechanisms of resistance and toxicity associated with platinating agents. *Cancer Treat. Rev.* **2007**, *33*, 9–23. [CrossRef]
38. Galluzzi, L.; Senovilla, L.; Vitale, I.; Michels, J.; Martins, I.; Kepp, O.; Castedo, M.; Kroemer, G. Molecular mechanisms of cisplatin resistance. *Oncogene* **2011**, *31*, 1869. [CrossRef]
39. Porchia, M.; Dolmella, A.; Gandin, V.; Marzano, C.; Pellei, M.; Peruzzo, V.; Refosco, F.; Santini, C.; Tisato, F. Neutral and charged phosphine/scorpionate copper(I) complexes: Effects of ligand assembly on their antiproliferative activity. *Eur. J. Med. Chem.* **2013**, *59*, 218–226. [CrossRef]
40. Tisato, F.; Marzano, C.; Porchia, M.; Pellei, M.; Santini, C. Copper in diseases and treatments, and copper-based anticancer strategies. *Med. Res. Rev.* **2009**, *30*, 708–749. [CrossRef]
41. De Feo, C.J.; Aller, S.G.; Siluvai, G.S.; Blackburn, N.J.; Unger, V.M. Three-dimensional structure of the human copper transporter hCTR1. *Proc. Natl. Acad. Sci. USA* **2009**, *106*, 4237. [CrossRef]
42. Puig, S.; Thiele, D.J. Molecular mechanisms of copper uptake and distribution. *Curr. Opin. Chem. Biol.* **2002**, *6*, 171–180. [CrossRef]
43. Puig, S.; Lee, J.; Lau, M.; Thiele, D.J. Biochemical and Genetic Analyses of Yeast and Human High Affinity Copper Transporters Suggest a Conserved Mechanism for Copper Uptake. *J. Biol. Chem.* **2002**, *277*, 26021–26030. [CrossRef]
44. Kaplan, J.H.; Maryon, E.B. How Mammalian Cells Acquire Copper: An Essential but Potentially Toxic Metal. *Biophys. J.* **2016**, *110*, 7–13. [CrossRef]
45. Porchia, M.; Benetollo, F.; Refosco, F.; Tisato, F.; Marzano, C.; Gandin, V. Synthesis and structural characterization of copper(I) complexes bearing N-methyl-1,3,5-triaza-7-phosphaadamantane (mPTA): Cytotoxic activity evaluation of a series of water soluble Cu(I) derivatives containing PTA, PTAH and mPTA ligands. *J. Inorg. Biochem.* **2009**, *103*, 1644–1651. [CrossRef]
46. Santini, C.; Pellei, M.; Papini, G.; Morresi, B.; Galassi, R.; Ricci, S.; Tisato, F.; Porchia, M.; Rigobello, M.P.; Gandin, V.; et al. In vitro antitumour activity of water soluble Cu(I), Ag(I) and Au(I) complexes supported by hydrophilic alkyl phosphine ligands. *J. Inorg. Biochem.* **2011**, *105*, 232–240. [CrossRef]
47. Marzano, C.; Gandin, V.; Pellei, M.; Colavito, D.; Papini, G.; Lobbia, G.G.; Del Giudice, E.; Porchia, M.; Tisato, F.; Santini, C. In Vitro Antitumor Activity of the Water Soluble Copper(I) Complexes Bearing the Tris(hydroxymethyl)phosphine Ligand. *J. Med. Chem.* **2008**, *51*, 798–808. [CrossRef]
48. Gandin, V.; Ceresa, C.; Esposito, G.; Indraccolo, S.; Porchia, M.; Tisato, F.; Santini, C.; Pellei, M.; Marzano, C. Therapeutic potential of the phosphino Cu(I) complex (HydroCuP) in the treatment of solid tumors. *Sci. Rep.* **2017**, *7*, 13936. [CrossRef]
49. Tisato, F.; Marzano, C.; Peruzzo, V.; Tegoni, M.; Giorgetti, M.; Damjanovic, M.; Trapananti, A.; Bagno, A.; Santini, C.; Pellei, M.; et al. Insights into the cytotoxic activity of the phosphane copper(I) complex [Cu(thp)4][PF6]. *J. Inorg. Biochem.* **2016**, *165*, 80–91. [CrossRef]

50. Gandin, V.; Tisato, F.; Dolmella, A.; Pellei, M.; Santini, C.; Giorgetti, M.; Marzano, C.; Porchia, M. In Vitro and in Vivo Anticancer Activity of Copper(I) Complexes with Homoscorpionate Tridentate Tris(pyrazolyl)borate and Auxiliary Monodentate Phosphine Ligands. *J. Med. Chem.* **2014**, *57*, 4745–4760. [CrossRef]
51. Venditti, I.; D'Amato, R.; Russo, M.V.; Falconieri, M. Synthesis of conjugated polymeric nanobeads for photonic bandgap materials. *Sens. Actuators B* **2007**, *126*, 35–40. [CrossRef]
52. De Angelis, R.; Venditti, I.; Fratoddi, I.; De Matteis, F.; Prosposito, P.; Cacciotti, I.; D'Amico, L.; Nanni, F.; Yadav, A.; Casalboni, M.; et al. From nanospheres to microribbons: Self-assembled Eosin Y doped PMMA nanoparticles as photonic crystals. *J. Colloid Interf. Sci.* **2014**, *414*, 24–32. [CrossRef]
53. Zangrando, M.; Zacchigria, M.; Finazzi, M.; Cocco, D.; Rochow, R.; Parmigiani, F. Polarized high-brilliance and high-resolution soft x-ray source at ELETTRA: The performance of beamline BACH. *Rev. Sci. Instrum.* **2004**, *75*, 31–36. [CrossRef]
54. Moulder, J.F.; Stickle, W.F.; Sobol, P.E.; Bomben, K.D. *Handbook of X-Ray Photoelectron Spectroscopy*; Perkin-Elmer Corporation: Eden Prairie, MN, USA, 1996.
55. Beamson, G.; Briggs, D. *High Resolution XPS of Organic Polymers*; John Wiley & Sons: Chichester, UK, 1992.
56. NIST X-ray Photoelectron Spectroscopy Database, Version 4.1 (National Institute of Standards and Technology). Available online: http://srdata.nist.gov/xps/.=NIST (accessed on 2 May 2019).
57. Nakayama, T.; Inamura, K.; Inoue, Y.; Ikeda, S.; Kishi, K. Adsorption of benzonitrile and alkyl cyanides on evaporated nickel and palladium films studied by XPS. *Surf. Sci.* **1987**, *179*, 47–58. [CrossRef]

nanomaterials

MDPI

Article

Shape-Controlled Synthesis of Au Nanostructures Using EDTA Tetrasodium Salt and Their Photothermal Therapy Applications

Youngjin Jang [1], Nohyun Lee [2], Jeong Hyun Kim [3], Yong Il Park [4,*] and Yuanzhe Piao [5,*]

1. Schulich Faculty of Chemistry, Technion-Israel Institute of Technology, Haifa 3200003, Israel; youngjin@technion.ac.il
2. School of Advanced Materials Engineering, Kookmin University, Seoul 02707, Korea; nohyunlee@kookmin.ac.kr
3. Center for Nanoparticle Research, Institute for Basic Science, and School of Chemical and Biological Engineering, Seoul National University, Seoul 151-742, Korea; jhkim113@snu.ac.kr
4. School of Chemical Engineering, Chonnam National University, Gwangju 61186, Korea
5. Graduate School of Convergence Science and Technology & Advanced Institutes of Convergence Technology, Seoul National University, Suwon 16229, Korea

* Correspondence: ypark@jnu.ac.kr (Y.I.P.); parkat9@snu.ac.kr (Y.P.); Tel.: +82-62-530-1886 (Y.I.P.); +82-31-888-9141 (Y.P.)

Received: 27 March 2018; Accepted: 16 April 2018; Published: 18 April 2018

Abstract: Tuning the optical properties of Au nanostructures is of paramount importance for scientific interest and has a wide variety of applications. Since the surface plasmon resonance properties of Au nanostructures can be readily adjusted by changing their shape, many approaches for preparing Au nanostructures with various shapes have been reported to date. However, complicated steps or the addition of several reagents would be required to achieve shape control of Au nanostructures. The present work describes a facile and effective shape-controlled synthesis of Au nanostructures and their photothermal therapy applications. The preparation procedure involved the reaction of $HAuCl_4$ and ethylenediaminetetraacetic acid (EDTA) tetrasodium salt, which acted as a reducing agent and ligand, at room temperature without the need for any toxic reagent or additives. The morphology control from spheres to branched forms and nanowire networks was easily achieved by varying the EDTA concentration. Detailed investigations revealed that the four carboxylic groups of the EDTA tetrasodium salt are essential for effective growth and stabilization. The produced Au nanowire networks exhibited a broad absorption band in the near-infrared (NIR) region, thereby showing efficient cancer therapeutic performance by inducing the selective photothermal destruction of cancerous glioblastoma cells (U87MG) under NIR irradiation.

Keywords: gold; nanostructure; EDTA tetrasodium salt; photothermal therapy

1. Introduction

To date, metal nanostructures have attracted a great deal of attention because of their intriguing electronic, optical, and catalytic properties [1–7]. In the past few decades, Au nanostructures of various shapes, including nanospheres [8–10], nanorods [11–19], nanowires [20–26], polyhedrons [27–32], nanoplates [33–35], nanoshells [36–38], and branched forms [39–49], have been extensively explored because of their shape-dependent surface plasmon resonance properties [50–53]. The fascinating surface plasmon resonance properties of Au nanostructures enable their effective implementation in a wide variety of biomedical applications, such as cancer therapy, bio-imaging, biological sensing, and diagnostics [54–58]. Biological targets (e.g., protein and DNA) are recognized as changes in the absorption of the functionalized Au nanostructures and an efficient bio-imaging using Au nanostructures is achieved by their distinctive interactions with light. Photothermal therapy is one of the major therapeutic methods that requires near-infrared (NIR) light, which shows maximum penetration depths in tissues and has a low toxicity for normal cells, so that cancer cells can be killed by heat generation [56,59–70]. To obtain Au nanostructures exhibiting optical properties active in the NIR region, intricate methods including multiple steps, many reagents, or long reaction times have been applied. For example, a seed-mediated growth approach has been widely used to produce anisotropic Au nanostructures, such as nanorods, which require steps for the formation of Au seeds and their growth. In addition, a method using two (or more) appropriate capping reagents with different binding affinities has been developed to obtain kinetic control of the growth rates on various crystal planes, resulting in the formation of non-spherical Au nanostructures. Therefore, the development of Au nanostructure preparation that yields NIR absorption is of prime importance for the successful extension to practical applications.

Recently, it was reported that ethylenediaminetetraacetic acid (EDTA), a well-known metal chelating ligand, can act as a reducing agent [71–73]. However, to the best of our knowledge, the morphology control of Au nanostructures by applying EDTA agents has not been attempted. Therefore, this is pioneering work on the preparation of Au nanocrystals exhibiting NIR activities using EDTA agents at room temperature.

Herein we present a facile and effective shape-controlled synthesis of colloidal Au nanostructures at room temperature and their effective performance for photothermal therapy. This simple procedure was achieved by reacting $HAuCl_4$ and EDTA tetrasodium salt, exhibiting a dual function as a reductant and a stabilizing agent, without any toxic compounds such as cetyltriethylammonium bromide or additional agents (e.g., superhydride). The morphology of the Au nanostructures was readily controlled by adjusting the molar ratio of the EDTA tetrasodium salt to the $HAuCl_4$ precursor. Reducing the concentration of the EDTA tetrasodium salt led to the formation of Au nanowire networks that are active in the NIR regime. Control experiments using several kinds of EDTA agents revealed that the four deprotonated carboxylic groups of EDTA tetrasodium salt play a key role as stabilizing agents and in effective growth control. The Au nanowire networks showed an effective and selective photothermal therapeutic effect on cancerous glioblastoma cells (U87MG) under NIR irradiation at 980 nm.

2. Materials and Methods

2.1. Materials

EDTA, EDTA disodium salt, and EDTA tetrasodium salt were purchased from Samchun Chemicals (Seoul, Korea). Tetrachloroaurate trihydrate ($HAuCl_4{\cdot}3H_2O$) was purchased from Strem Chemicals, Inc. (Newburyport, MA, USA). Methoxy poly(ethylene glycol)sulfhydryl (mPEG-SH, M_w = 5000) was purchased from SunBio Corp. (Anyang, Korea). Water deionized by a Nano Pure System (Barnstead, Thermo Fisher Scientific, Waltham, MA, USA) was used. The chemicals used for the preparation of the solutions were purchased at the highest grade possible, and used without further purification.

2.2. Shape-Controlled Synthesis of Au Nanostructures

To synthesize Au nanowires with a network structure, 15 mg of EDTA tetrasodium salt was injected into 7 mL of deionized water at room temperature, and the mixture solution was stirred for a few minutes to ensure complete dissolution of the salt. Further, 0.1 mL of 0.1 M $HAuCl_4$ was added to the aqueous solution containing the EDTA tetrasodium salt at room temperature, and the resulting solution was vigorously stirred for 1 h. After the reaction, the mixture was washed with water by centrifugation at 14,000 rpm for 10 min. To synthesize the branched Au nanoparticles, the molar ratio of $HAuCl_4$ to EDTA tetrasodium salt was changed to 1:6, and other conditions were kept constant. To synthesize the spherical Au nanoparticles, the molar ratio of $HAuCl_4$ to EDTA tetrasodium salt was increased to 1:8.

To stabilize the Au nanowire networks, an mPEG-SH aqueous solution prepared by dissolving 40 mg of mPEG-SH in 2 mL of deionized water was injected into the mixture solution and further stirred for 30 min. The resulting solution was washed with water by centrifugation at 14,000 rpm for 15 min.

2.3. *In Vitro Cytotoxicity against U87MG Cells*

U87MG human glioblastoma cells were grown as monolayer cultures in a 100-mm dish and subcultured 3 times a week at 37 °C in an atmosphere of 5% CO_2 and 100% relative humidity. For the in vitro cytotoxicity assay, cells at a logarithmic growth phase were detached and plated (0.2 mL per well) in 96-well flat-bottomed microplates at a density of 10,000 cells per well, which were then left for 1–2 days at 37 °C to resume exponential growth. After washing the cells with phosphate buffered saline (PBS), 0.1 mL of the culture medium, with various concentrations of Au nanowire networks coated with mPEG-SH ligand, was added to the wells in quintuplicate. For the control wells, the same volume of culture medium was included in each experiment. Following 24 h of continuous exposure to the Au nanowire networks under 5% CO_2 atmosphere at 37 °C, cell survival was assessed using the (3-(4,5-dimethylthiazol-2-yl)-2,5-diphenyltetrazolium bromide (MTT) cell proliferation assay.

2.4. *In Vitro Photothermal Therapy*

The U87MG human glioblastoma cells were plated in a confocal dish and allowed to grow on the coverslip. Additionally, 150 μg Au/mL of the Au nanowire networks coated with mPEG-SH ligand were incubated with the cells for 24 h. The unbound Au nanowire networks were rinsed with the PBS buffer, and the cells were immersed in the culture medium.

Subsequently, a 980-nm continuous wave (CW) diode laser was used for irradiation and was focused to a 1-mm-diameter spot on the sample (power density: 38.2 W/cm^2). The cells were irradiated for 10 min. The control cells without the Au nanowire networks with laser irradiation were also tested. Subsequently, the cells were stained with a few drops of 0.4% trypan blue to test viability. The dead cells were stained blue, while the live cells remained clear. After staining, the cells were rinsed with the PBS buffer and immersed in the culture medium for bright-field imaging.

2.5. *Characterization*

Transmission electron microscopy (TEM) and high-resolution TEM (HR-TEM) images were acquired using a JEOL EM-2010 microscope (JEOL, Tokyo, Japan) at an accelerating voltage of 200 kV. The powder X-ray diffraction (XRD) pattern was obtained by a Rigaku D/Max-3C diffractometer (Cu Kα radiation, λ = 0.15418 nm, Rigaku Co. Ltd., Tokyo, Japan). The UV-Vis absorption spectra were acquired by a Jasco V-570-type spectrophotometer (Jasco SLM-468, Tokyo, Japan). Elemental analysis was performed by inductively coupled plasma-atomic emission spectroscopy (ICP-AES) using an ICPS-7500 spectrometer (Shimadzu, Kyoto, Japan). The MTT cell proliferation assay was performed by an ELx808TM absorbance microplate reader (Biotek Instruments Inc., Winooski, VT, USA).

3. Results and Discussion

In general, colloidal synthesis of Au nanostructures requires reductants and capping agents. In the current procedure, the Au nanostructures were readily synthesized by reacting $HAuCl_4$ and EDTA tetrasodium salt in an aqueous solution at room temperature. The EDTA tetrasodium salt played a dual role as a reducing agent and as a ligand, thus allowing for simple synthesis without any additives. After the reaction, the products were washed with water to remove any byproducts and unreacted reagents. More details are provided in the experimental section.

Figure 1a shows a representative TEM image of the Au nanostructures prepared using a 1:8 molar ratio of $HAuCl_4$ to EDTA tetrasodium salt. The image presents the formation of spherical Au nanocrystals with an average diameter of 11 nm. A HR-TEM image is shown in the inset of Figure 1a, indicating the polycrystallinity of the Au nanocrystals with pronounced lattice fringes of 0.235 nm, corresponding to the (111) planes of a face-centered-cubic (fcc) Au crystal structure. Decreasing the molar ratio of $HAuCl_4$ to EDTA tetrasodium salt induced shape control, resulting in branched forms (see Figure 1b) at 1:6 and nanowire networks (see Figure 1c) at 1:4. When the molar ratio was below 1:3, severely aggregated Au particles were observed because of the insufficient capping agent. Based on these observations, the formation of the Au nanostructures is summarized in Scheme 1, illustrating that the morphology of the Au nanostructures is easily controlled by adjusting the EDTA tetrasodium salt concentration. High concentrations of EDTA tetrasodium salt produced only spherical Au nanocrystals. Decreasing the molar ratio of $HAuCl_4$ to EDTA tetrasodium salt resulted in the formation of non-spherical Au nanostructures, such as branched forms and nanowire networks.

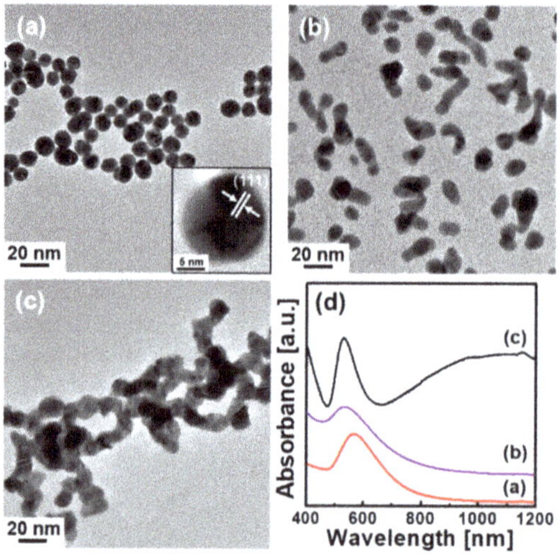

Figure 1. Transmission electron microscopy (TEM) images of Au nanostructures prepared using different molar ratios of $HAuCl_4$ to ethylenediaminetetraacetic acid (EDTA) tetrasodium salt: (**a**) 1:8, the inset is the high-resolution TEM (HR-TEM) image of panel (**a**); (**b**) 1:6; (**c**) 1:4; (**d**) Absorption spectra of Au nanostructures in panels (**a**–**c**).

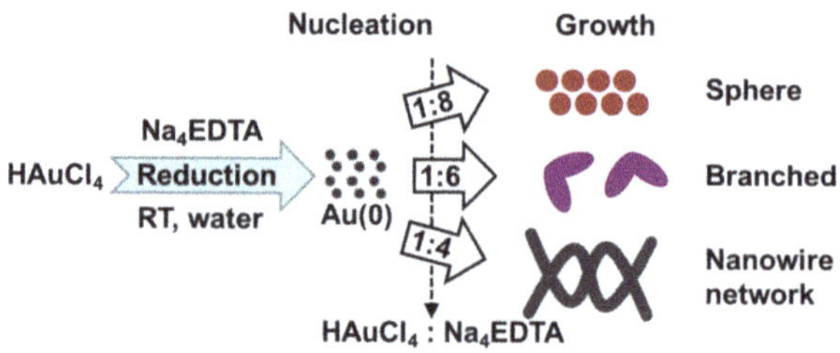

Scheme 1. Schematic illustration of morphology control by adjusting the molar ratio of $HAuCl_4$ to EDTA tetrasodium salt.

The absorption spectra of Au nanostructures, with three kinds of morphologies, are presented in Figure 1d, clearly displaying the shape-dependent surface plasmon resonance behaviors. Spherical Au nanocrystals (see red curve (a) in Figure 1d) exhibit one absorption peak at 570 nm, resulting from the transverse mode, while the transverse band becomes broader in the branched Au nanostructures (see purple curve (b) in Figure 1d). The most intriguing activity was observed when using a 1:4 molar ratio of $HAuCl_4$ to EDTA tetrasodium salt (see black curve (c) in Figure 1d). Two distinct optical behaviors were identified in the absorption spectrum of the Au nanowire networks. The peak at 530 nm is attributed to transverse surface plasmon resonance (TSPR), and a broad absorption band over the NIR spectra regime was observed, owing to the longitudinal surface plasmon resonance (LSPR).

The Au nanowire networks showing absorption in the NIR regime were further investigated, as described below. The low-magnification TEM image of the Au nanowire networks in Figure 2a reveals that the wires have considerable lengths of a few micrometers. HR-TEM images of the Au nanowire networks are displayed in Figure 2b,c, indicating the dominant (111) lattice planes, in agreement with the fast Fourier transform (FFT) image shown in the inset. Moreover, these images provide insight into the crystallinity of the Au nanowire networks and their growth characteristics, acquired by the coalescence of Au particles and subsequent deposition of Au atoms, as suggested in previous studies [21,22,74].

The XRD pattern of the Au nanowire networks is presented in Figure 2d, showing that all characteristic peaks are indexed to the (111), (200), (220), (311) and (222) planes of the fcc Au crystal structure without any other phases, which also supports the TEM analysis.

When the solution containing the isolated Au nanowire networks was kept at room temperature for approximately one day, the color gradually changed from black to deep red, implying the transformation of the Au nanowire networks into spherical Au nanoparticles. The shape instability of nonspherical Au nanostructures has also been observed in previous works [21,22,42,43] and explained by the thermodynamic instability of the nonspheres [42,43]. To prevent the transformation of the Au nanowire networks, a stabilizing agent exhibiting stronger binding, the mPEG-SH ligand, was used. mPEG-SH stabilized the Au nanowire networks and remained intact without any color change over one month, as shown in Figure S1 in the Supplementary Materials. The absorption spectra and TEM image in Figure S1 also support that mPEG-SH coating stabilize Au nanowire networks without morphological change.

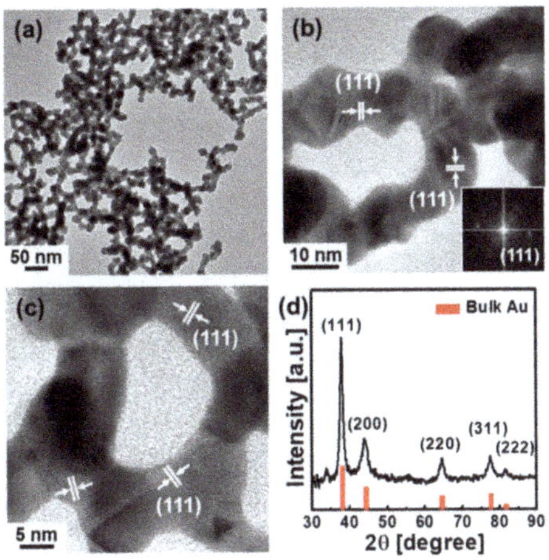

Figure 2. Characteristics of Au nanowire networks: (**a**,**b**) TEM images with different magnifications, the inset is the fast Fourier transform (FFT) image obtained from the panel (b); (**c**) HR-TEM images; (**d**) X-ray diffraction (XRD) pattern, red vertical lines indicate peak positions and relative intensities of bulk Au.

To monitor the temporal evolution of the Au nanowire networks during the reaction time of 1 h, aliquots withdrawn at different reaction stages were analyzed, as shown in Figure 3. As seen in the TEM images, the length and diameter of the wires gradually increased as the reaction time progressed, implying that the Au nanowire networks were formed by depositing the Au monomer at the joints between the assembled Au nanocrystals. The absorption spectra of aliquots taken at different reaction stages are also shown in Figure S2, indicating that only one broad feature was seen at the beginning stages (e.g., 15 min). After 25 min, two distinct peaks were clearly observed, due to the TSPR and LSPR modes. This evolution shows that Au nanowire networks formed in nearly 25 min, in good agreement with the TEM analysis (see Figure 3f).

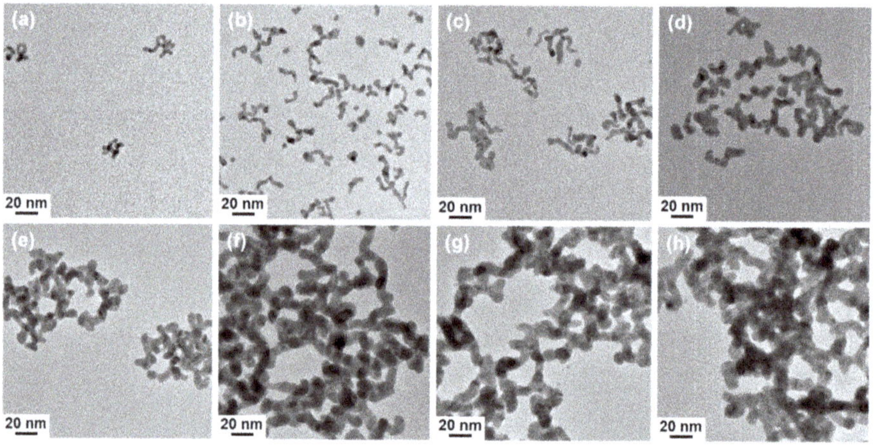

Figure 3. TEM images of the samples at different stages of the reaction: (**a**) 2 min; (**b**) 5 min; (**c**) 10 min; (**d**) 15 min; (**e**) 20 min; (**f**) 25 min; (**g**) 45 min; (**h**) 60 min.

Several kinds of control experiments were performed to understand the effects of synthetic conditions. Without the addition of the EDTA tetrasodium salt, no Au nanostructures were formed. This supports the fact that the four carboxylic groups of the EDTA tetrasodium salt reduce the Au^{3+} species, which is in agreement with previous works [71–73]. It is noteworthy that EDTA tetrasodium salt introduced strong basic conditions, thus eliminating the need for additional steps (see Table S1 in the Supplementary Materials), whereas adjusting the pH by adding a suitable base was necessary for the preparation of the Au nanostructures in previous studies [72,75,76].

In addition, the addition of other EDTA salts was attempted for the synthesis of Au nanostructures. Under similar conditions such as temperature and concentration, the use of EDTA and EDTA disodium salt led to the formation of micrometer-sized and submicrometer-sized Au aggregates, respectively, as shown in Figure S3, revealing that the four deprotonated carboxyl groups (COO^-) of EDTA tetrasodium salt as a tetradentate group are essential for allowing effective growth on the nanoscale.

NIR absorption is crucial for many biomedical applications because biological tissues, blood, and water show low absorption in this wavelength range [77,78]. The cytotoxicity of the Au nanowire networks coated with mPEG-SH was evaluated to demonstrate their applicability as photothermal therapeutic agents. The MTT cell proliferation assay is presented in Figure S4, indicating that more than 80% of the U87MG cells survived up to 1.27 mM (250 μg/mL) of the Au concentration. A more detailed procedure is presented in the Experimental Section.

Figure 4 represents the images of the U87MG cells without and with the Au nanowire networks after the NIR laser irradiation at 980 nm for 10 min, clearly revealing cell damage at the center of the laser beam for cells with an incubation of the Au nanowire networks, while the control cells remained intact. The dead cells were stained blue by 0.4% trypan blue. This observation clearly demonstrates the effectiveness of the Au nanowire networks in NIR photothermal therapy by selective killing of cancer cells, because of the local heating generated by their effective NIR absorption. Because of their relatively large size and irregular shape, the Au nanowire networks may not be suitable for animal study by systemic delivery. However, due to the good photothermal effect by NIR light, they can be used as a local heat generator to regulate cellular activity by attaching them to the cell membrane or biochip surface [79].

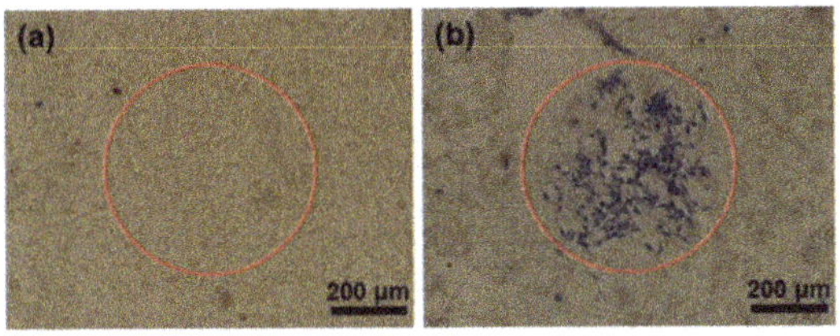

Figure 4. Optical microscope images of cancerous glioblastoma cells (U87MG) cells incubated (**a**) without; (**b**) with the Au nanowire networks, after irradiation for 10 min with a 980-nm near-infrared (NIR) CW diode laser and subsequent staining with 0.4% trypan blue.

4. Conclusions

In summary, our work presents a facile and effective shape-controlled synthesis of Au nanostructures and their photothermal therapeutic effect. The described procedure involves the simple mixing of $HAuCl_4$ and EDTA tetrasodium salt in an aqueous solution at room temperature, without additional ligands or toxic reagents. Adjusting the molar ratios of $HAuCl_4$ to EDTA tetrasodium salt enables effective morphology control of Au nanostructures from spheres to branched forms and nanowire networks. Detailed control experiments revealed that the four deprotonated carboxylic acids of the EDTA tetrasodium salt provided effective growth control and stabilization. The Au nanowire networks showed strong absorption in the NIR region and hence were suitable for photothermal therapy. Under NIR irradiation, the Au nanowire networks allowed for selective destruction of cancerous U87MG cells by local heating, generated by the NIR absorption. This work demonstrates the development of a simple synthetic route to NIR-active Au nanostructures, which can be extended to other applications including optical sensing and surface-enhanced Raman scattering.

Supplementary Materials: The following are available online at http://www.mdpi.com/2079-4991/8/4/252/s1, Figure S1: (a) Photograph of mPEG-SH stabilized Au nanowire networks kept in ambient conditions for one month, showing good stability, (b) absorption spectra of as-synthesized Au nanowire networks (black curve) and mPEG-SH coated Au nanowire networks (red curve), and (c) TEM image of Au nanowire networks after mPEG-SH coating, Figure S2: Absorption spectra of Au nanostructures taken at different stages of the reaction, Figure S3: TEM images of Au aggregates generated by using (a) EDTA and (b) EDTA disodium salt, Figure S4: The cell viability of Au nanowire networks by MTT assay, Table S1: The pH values at different reaction conditions.

Acknowledgments: Y.P. was supported by the Center for Integrated Smart Sensors, funded by the Ministry of Science, ICT and Future Planning, Republic of Korea, as a Global Frontier Project (CISS-012M3A6A6054186). Y.I.P. was supported by the National Research Foundation of Korea (NRF) grant funded by the Korean government (Ministry of Science, ICT & Future Planning) (No. 2016R1A4A1012224).

Author Contributions: Y.J., Y.I.P. and Y.P. conceived and designed the experiments; Y.J. and Y.P. performed the synthesis experiments; Y.I.P. performed the photothermal therapy experiments; N.L. performed the MTT assay experiments; J.H.K. executed the article editing; Y.J., Y.I.P. and Y.P. wrote the paper.

Conflicts of Interest: The authors declare no conflict of interest.

References

1. Schmid, G.; Corain, B. Nanoparticulated gold: Syntheses, structures, electronics, and reactivities. *Eur. J. Inorg. Chem.* **2003**, *2003*, 3081–3098. [CrossRef]
2. Daniel, M.C.; Astruc, D. Gold nanoparticles: Assembly, supramolecular chemistry, quantum-size-related properties, and applications toward biology, catalysis, and nanotechnology. *Chem. Rev.* **2004**, *104*, 293–346. [CrossRef] [PubMed]

3. Eustis, S.; El-Sayed, M.A. Why gold nanoparticles are more precious than pretty gold: Noble metal surface plasmon resonance and its enhancement of the radiative and nonradiative properties of nanocrystals of different shapes. *Chem. Soc. Rev.* **2006**, *35*, 209–217. [CrossRef] [PubMed]
4. Hutchings, G.J.; Brust, M.; Schmidbaur, H. Gold–an introductory perspective. *Chem. Soc. Rev.* **2008**, *37*, 1759–1765. [CrossRef] [PubMed]
5. Grzelczak, M.; Perez-Juste, J.; Mulvaney, P.; Liz-Marzan, L.M. Shape control in gold nanoparticle synthesis. *Chem. Soc. Rev.* **2008**, *37*, 1783–1791. [CrossRef] [PubMed]
6. Zhang, Y.; Chu, W.; Foroushani, A.D.; Wang, H.; Li, D.; Liu, J.; Barrow, C.J.; Wang, X.; Yang, W. New gold nanostructures for sensor applications: A review. *Materials* **2014**, *7*, 5169–5201. [CrossRef] [PubMed]
7. Salunke, B.K.; Sathiyamoorthi, E.; Tran, T.K.; Kim, B.S. Phyto-synthesized silver nanoparticles for biological applications. *Korean J. Chem. Eng.* **2017**, *34*, 943–951. [CrossRef]
8. Brust, M.; Walker, M.; Bethell, D.; Schiffrin, D.J.; Whyman, R. Synthesis of thiol-derivatised gold nanoparticles in a two-phase liquid–liquid system. *J. Chem. Soc. Chem. Commun.* **1994**, *0*, 801–802. [CrossRef]
9. Kim, S.; Jang, Y.; Yoon, K.Y.; Park, J. Surface engineered gold nanoparticles through highly stable metal-surfactant complexes. *J. Colloid Interface Sci.* **2016**, *464*, 110–116. [CrossRef] [PubMed]
10. Kim, I.; Lee, S.-H.; Lim, B.; Park, B.J.; Bhang, S.H.; Yu, T. Aqueous-phase synthesis of metal nanoparticles using phosphates as stabilizers. *Korean J. Chem. Eng.* **2017**, *34*, 231–233. [CrossRef]
11. Nikoobakht, B.; El-Sayed, M.A. Preparation and growth mechanism of gold nanorods (NRs) using seed-mediated growth method. *Chem. Mater.* **2003**, *15*, 1957–1962. [CrossRef]
12. Murphy, C.J.; Jana, N.R. Controlling the aspect ratio of inorganic nanorods and nanowires. *Adv. Mater.* **2002**, *14*, 80–82. [CrossRef]
13. Jana, N.R.; Gearheart, L.; Murphy, C.J. Seed-mediated growth approach for shape-controlled synthesis of spheroidal and rod-like gold nanoparticles using a surfactant template. *Adv. Mater.* **2001**, *13*, 1389–1393. [CrossRef]
14. Kim, F.; Song, J.H.; Yang, P. Photochemical synthesis of gold nanorods. *J. Am. Chem. Soc.* **2002**, *124*, 14316–14317. [CrossRef] [PubMed]
15. Pérez-Juste, J.; Pastoriza-Santos, I.; Liz-Marzán, L.M.; Mulvaney, P. Gold nanorods: Synthesis, characterization and applications. *Coord. Chem. Rev.* **2005**, *249*, 1870–1901. [CrossRef]
16. Chang, H.-H.; Murphy, C.J. Mini gold nanorods with tunable plasmonic peaks beyond 1000 nm. *Chem. Mater.* **2018**, *30*, 1427–1435. [CrossRef]
17. Zhang, L.; Xia, K.; Lu, Z.; Li, G.; Chen, J.; Deng, Y.; Li, S.; Zhou, F.; He, N. Efficient and facile synthesis of gold nanorods with finely tunable plasmonic peaks from visible to near-IR range. *Chem. Mater.* **2014**, *26*, 1794–1798. [CrossRef]
18. Zhang, Q.; Zhou, Y.; Villarreal, E.; Lin, Y.; Zou, S.; Wang, H. Faceted gold nanorods: Nanocuboids, convex nanocuboids, and concave nanocuboids. *Nano Lett.* **2015**, *15*, 4161–4169. [CrossRef] [PubMed]
19. Wang, Y.N.; Wei, W.T.; Yang, C.W.; Huang, M.H. Seed-mediated growth of ultralong gold nanorods and nanowires with a wide range of length tunability. *Langmuir* **2013**, *29*, 10491–10497. [CrossRef] [PubMed]
20. Lu, X.; Yavuz, M.S.; Tuan, H.Y.; Korgel, B.A.; Xia, Y. Ultrathin gold nanowires can be obtained by reducing polymeric strands of oleylamine-aucl complexes formed via aurophilic interaction. *J. Am. Chem. Soc.* **2008**, *130*, 8900–8901. [CrossRef] [PubMed]
21. Pong, B.-K.; Elim, H.I.; Chong, J.-X.; Ji, W.; Trout, B.L.; Lee, J.-Y. New insights on the nanoparticle growth mechanism in the citrate reduction of gold(III) salt: Formation of the Au nanowire intermediate and its nonlinear optical properties. *J. Phys. Chem. C* **2007**, *111*, 6281–6287. [CrossRef]
22. Pei, L.; Mori, K.; Adachi, M. Formation process of two-dimensional networked gold nanowires by citrate reduction of $AuCl^{4-}$ and the shape stabilization. *Langmuir* **2004**, *20*, 7837–7843. [CrossRef] [PubMed]
23. Navaladian, S.; Janet, C.M.; Viswanathan, B.; Varadarajan, T.K.; Viswanath, R.P. A facile room-temperature synthesis of gold nanowires by oxalate reduction method. *J. Phys. Chem. C* **2007**, *111*, 14150–14156. [CrossRef]
24. Huo, Z.; Tsung, C.K.; Huang, W.; Zhang, X.; Yang, P. Sub-two nanometer single crystal Au nanowires. *Nano Lett.* **2008**, *8*, 2041–2044. [CrossRef] [PubMed]
25. Zhu, C.; Peng, H.C.; Zeng, J.; Liu, J.; Gu, Z.; Xia, Y. Facile synthesis of gold wavy nanowires and investigation of their growth mechanism. *J. Am. Chem. Soc.* **2012**, *134*, 20234–20237. [CrossRef] [PubMed]
26. Chirea, M.; Freitas, A.; Vasile, B.S.; Ghitulica, C.; Pereira, C.M.; Silva, F. Gold nanowire networks: Synthesis, characterization, and catalytic activity. *Langmuir* **2011**, *27*, 3906–3913. [CrossRef] [PubMed]

27. Seo, D.; Park, J.C.; Song, H. Polyhedral gold nanocrystals with O h symmetry: From octahedra to cubes. *J. Am. Chem. Soc.* **2006**, *128*, 14863–14870. [CrossRef] [PubMed]
28. Seo, D.; Yoo, C.I.; Park, J.C.; Park, S.M.; Ryu, S.; Song, H. Directed surface overgrowth and morphology control of polyhedral gold nanocrystals. *Angew. Chem. Int. Ed.* **2008**, *47*, 763–767. [CrossRef] [PubMed]
29. Heo, J.; Kim, D.S.; Kim, Z.H.; Lee, Y.W.; Kim, D.; Kim, M.; Kwon, K.; Park, H.J.; Yun, W.S.; Han, S.W. Controlled synthesis and characterization of the enhanced local field of octahedral Au nanocrystals. *Chem. Commun.* **2008**, *0*, 6120–6122. [CrossRef] [PubMed]
30. Sun, Y.; Xia, Y. Shape-controlled synthesis of gold and silver nanoparticles. *Science* **2002**, *298*, 2176–2179. [CrossRef] [PubMed]
31. Sau, T.K.; Murphy, C.J. Room temperature, high-yield synthesis of multiple shapes of gold nanoparticles in aqueous solution. *J. Am. Chem. Soc.* **2004**, *126*, 8648–8649. [CrossRef] [PubMed]
32. Lee, J.H.; Gibson, K.J.; Chen, G.; Weizmann, Y. Bipyramid-templated synthesis of monodisperse anisotropic gold nanocrystals. *Nat. Commun.* **2015**, *6*, 7571. [CrossRef] [PubMed]
33. Ah, C.S.; Yun, Y.J.; Park, H.J.; Kim, W.-J.; Ha, D.H.; Yun, W.S. Size-controlled synthesis of machinable single crystalline gold nanoplates. *Chem. Mater.* **2005**, *17*, 5558–5561. [CrossRef]
34. Xin, W.; Severino, J.; De Rosa, I.M.; Yu, D.; McKay, J.; Ye, P.; Yin, X.; Yang, J.M.; Carlson, L.; Kodambaka, S. One-step synthesis of tunable-size gold nanoplates on graphene multilayers. *Nano Lett.* **2018**, *18*, 1875–1881. [CrossRef] [PubMed]
35. Huang, Y.; Ferhan, A.R.; Gao, Y.; Dandapat, A.; Kim, D.H. High-yield synthesis of triangular gold nanoplates with improved shape uniformity, tunable edge length and thickness. *Nanoscale* **2014**, *6*, 6496–6500. [CrossRef] [PubMed]
36. Oldenburg, S.J.; Averitt, R.D.; Westcott, S.L.; Halas, N.J. Nanoengineering of optical resonances. *Chem. Phys. Lett.* **1998**, *288*, 243–247. [CrossRef]
37. Graf, C.; van Blaaderen, A. Metallodielectric colloidal core−shell particles for photonic applications. *Langmuir* **2002**, *18*, 524–534. [CrossRef]
38. Gao, Y.; Gu, J.; Li, L.; Zhao, W.; Li, Y. Synthesis of gold nanoshells through improved seed-mediated growth approach: Brust-like, in situ seed formation. *Langmuir* **2016**, *32*, 2251–2258. [CrossRef] [PubMed]
39. Wu, H.-L.; Chen, C.-H.; Huang, M.H. Seed-mediated synthesis of branched gold nanocrystals derived from the side growth of pentagonal bipyramids and the formation of gold nanostars. *Chem. Mater.* **2009**, *21*, 110–114. [CrossRef]
40. Xie, J.; Lee, J.Y.; Wang, D.I.C. Seedless, surfactantless, high-yield synthesis of branched gold nanocrystals in hepes buffer solution. *Chem. Mater.* **2007**, *19*, 2823–2830. [CrossRef]
41. Li, Z.; Li, W.; Camargo, P.H.; Xia, Y. Facile synthesis of branched Au nanostructures by templating against a self-destructive lattice of magnetic Fe nanoparticles. *Angew. Chem. Int. Ed.* **2008**, *47*, 9653–9656. [CrossRef] [PubMed]
42. Kuo, C.H.; Huang, M.H. Synthesis of branched gold nanocrystals by a seeding growth approach. *Langmuir* **2005**, *21*, 2012–2016. [CrossRef] [PubMed]
43. Wu, H.Y.; Liu, M.; Huang, M.H. Direct synthesis of branched gold nanocrystals and their transformation into spherical nanoparticles. *J. Phys. Chem. B* **2006**, *110*, 19291–19294. [CrossRef] [PubMed]
44. Chen, S.; Wang, Z.L.; Ballato, J.; Foulger, S.H.; Carroll, D.L. Monopod, bipod, tripod, and tetrapod gold nanocrystals. *J. Am. Chem. Soc.* **2003**, *125*, 16186–16187. [CrossRef] [PubMed]
45. Gao, X.; Xing, G.; Chu, W.; Liang, X.; Zhao, Y.; Jing, L.; Yuan, H.; Cui, Y.; Dong, J. The growth of complex nanostructures: Synergism of dipolar force and stacking-defects in anisotropic self-assembly. *Adv. Mater.* **2008**, *20*, 1794–1798. [CrossRef]
46. Liao, H.G.; Jiang, Y.X.; Zhou, Z.Y.; Chen, S.P.; Sun, S.G. Shape-controlled synthesis of gold nanoparticles in deep eutectic solvents for studies of structure-functionality relationships in electrocatalysis. *Angew. Chem. Int. Ed.* **2008**, *47*, 9100–9103. [CrossRef] [PubMed]
47. Mohanty, A.; Garg, N.; Jin, R. A universal approach to the synthesis of noble metal nanodendrites and their catalytic properties. *Angew. Chem. Int. Ed.* **2010**, *49*, 4962–4966. [CrossRef] [PubMed]
48. Abalde-Cela, S.; Taladriz-Blanco, P.; de Oliveira, M.G.; Abell, C. Droplet microfluidics for the highly controlled synthesis of branched gold nanoparticles. *Sci. Rep.* **2018**, *8*, 2440. [CrossRef] [PubMed]

49. Sajitha, M.; Vindhyasarumi, A.; Gopi, A.; Yoosaf, K. Shape controlled synthesis of multi-branched gold nanocrystals through a facile one-pot bifunctional biomolecular approach. *RSC Adv.* **2015**, *5*, 98318–98324. [CrossRef]
50. Nehl, C.L.; Hafner, J.H. Shape-dependent plasmon resonances of gold nanoparticles. *J. Mater. Chem.* **2008**, *18*, 2415–2419. [CrossRef]
51. Jain, P.K.; Huang, X.; El-Sayed, I.H.; El-Sayed, M.A. Noble metals on the nanoscale: Optical and photothermal properties and some applications in imaging, sensing, biology, and medicine. *Acc. Chem. Res.* **2008**, *41*, 1578–1586. [CrossRef] [PubMed]
52. Myroshnychenko, V.; Rodriguez-Fernandez, J.; Pastoriza-Santos, I.; Funston, A.M.; Novo, C.; Mulvaney, P.; Liz-Marzan, L.M.; Garcia de Abajo, F.J. Modelling the optical response of gold nanoparticles. *Chem. Soc. Rev.* **2008**, *37*, 1792–1805. [CrossRef] [PubMed]
53. Wang, A.X.; Kong, X. Review of recent progress of plasmonic materials and nano-structures for surface-enhanced raman scattering. *Materials* **2015**, *8*, 3024–3052. [CrossRef] [PubMed]
54. Dreaden, E.C.; Mackey, M.A.; Huang, X.; Kang, B.; El-Sayed, M.A. Beating cancer in multiple ways using nanogold. *Chem. Soc. Rev.* **2011**, *40*, 3391–3404. [CrossRef] [PubMed]
55. Giljohann, D.A.; Seferos, D.S.; Daniel, W.L.; Massich, M.D.; Patel, P.C.; Mirkin, C.A. Gold nanoparticles for biology and medicine. *Angew. Chem. Int. Ed.* **2010**, *49*, 3280–3294. [CrossRef] [PubMed]
56. Lal, S.; Clare, S.E.; Halas, N.J. Nanoshell-enabled photothermal cancer therapy: Impending clinical impact. *Acc. Chem. Res.* **2008**, *41*, 1842–1851. [CrossRef] [PubMed]
57. Li, N.; Zhao, P.; Astruc, D. Anisotropic gold nanoparticles: Synthesis, properties, applications, and toxicity. *Angew. Chem. Int. Ed.* **2014**, *53*, 1756–1789. [CrossRef] [PubMed]
58. Park, H.; Whang, K.; Shin, Y.; Lee, J.; Kang, T. Synthesis of colloidal plasmonic microspheres via spontaneous formation and three-dimensional assembly of metal nanoparticles. *Korean J. Chem. Eng.* **2017**, *34*, 2086–2091. [CrossRef]
59. Kim, J.; Park, S.; Lee, J.E.; Jin, S.M.; Lee, J.H.; Lee, I.S.; Yang, I.; Kim, J.S.; Kim, S.K.; Cho, M.H.; et al. Designed fabrication of multifunctional magnetic gold nanoshells and their application to magnetic resonance imaging and photothermal therapy. *Angew. Chem. Int. Ed.* **2006**, *45*, 7754–7758. [CrossRef] [PubMed]
60. Huang, X.; Neretina, S.; El-Sayed, M.A. Gold nanorods: From synthesis and properties to biological and biomedical applications. *Adv. Mater.* **2009**, *21*, 4880–4910. [CrossRef] [PubMed]
61. Park, H.; Yang, J.; Lee, J.; Haam, S.; Choi, I.H.; Yoo, K.H. Multifunctional nanoparticles for combined doxorubicin and photothermal treatments. *ACS Nano* **2009**, *3*, 2919–2926. [CrossRef] [PubMed]
62. Nam, J.; Won, N.; Jin, H.; Chung, H.; Kim, S. Ph-induced aggregation of gold nanoparticles for photothermal cancer therapy. *J. Am. Chem. Soc.* **2009**, *131*, 13639–13645. [CrossRef] [PubMed]
63. Au, L.; Zheng, D.; Zhou, F.; Li, Z.Y.; Li, X.; Xia, Y. A quantitative study on the photothermal effect of immuno gold nanocages targeted to breast cancer cells. *ACS Nano* **2008**, *2*, 1645–1652. [CrossRef] [PubMed]
64. Yang, J.; Lee, J.; Kang, J.; Oh, S.J.; Ko, H.J.; Son, J.H.; Lee, K.; Suh, J.S.; Huh, Y.M.; Haam, S. Smart drug-loaded polymer gold nanoshells for systemic and localized therapy of human epithelial cancer. *Adv. Mater.* **2009**, *21*, 4339–4342. [CrossRef] [PubMed]
65. Li, J.L.; Day, D.; Gu, M. Ultra-low energy threshold for cancer photothermal therapy using transferrin-conjugated gold nanorods. *Adv. Mater.* **2008**, *20*, 3866–3871. [CrossRef]
66. Hu, K.W.; Liu, T.M.; Chung, K.Y.; Huang, K.S.; Hsieh, C.T.; Sun, C.K.; Yeh, C.S. Efficient near-IR hyperthermia and intense nonlinear optical imaging contrast on the gold nanorod-in-shell nanostructures. *J. Am. Chem. Soc.* **2009**, *131*, 14186–14187. [CrossRef] [PubMed]
67. Qiu, P.; Yang, M.; Qu, X.; Huai, Y.; Zhu, Y.; Mao, C. Tuning photothermal properties of gold nanodendrites for in vivo cancer therapy within a wide near infrared range by simply controlling their degree of branching. *Biomaterials* **2016**, *104*, 138–144. [CrossRef] [PubMed]
68. Fazal, S.; Jayasree, A.; Sasidharan, S.; Koyakutty, M.; Nair, S.V.; Menon, D. Green synthesis of anisotropic gold nanoparticles for photothermal therapy of cancer. *ACS Appl. Mater. Interfaces* **2014**, *6*, 8080–8089. [CrossRef] [PubMed]
69. Zhang, Q.; Wang, L.; Jiang, Y.; Gao, W.; Wang, Y.; Yang, X.; Yang, X.; Liu, Z. Gold nanorods with silica shell and pamam dendrimers for efficient photothermal therapy and low toxic codelivery of anticancer drug and sirna. *Adv. Mater. Interfaces* **2017**, *4*, 1701166. [CrossRef]

70. Mocan, L.; Matea, C.; Tabaran, F.A.; Mosteanu, O.; Pop, T.; Puia, C.; Agoston-Coldea, L.; Gonciar, D.; Kalman, E.; Zaharie, G.; et al. Selective in vitro photothermal nano-therapy of mrsa infections mediated by igg conjugated gold nanoparticles. *Sci. Rep.* **2016**, *6*, 39466. [CrossRef] [PubMed]
71. Guo, R.; Zhang, L.; Zhu, Z.; Jiang, X. Direct facile approach to the fabrication of chitosan-gold hybrid nanospheres. *Langmuir* **2008**, *24*, 3459–3464. [CrossRef] [PubMed]
72. Dozol, H.; Mériguet, G.; Ancian, B.; Cabuil, V.; Xu, H.; Wang, D.; Abou-Hassan, A. On the synthesis of Au nanoparticles using edta as a reducing agent. *J. Phys. Chem. C* **2013**, *117*, 20958–20966. [CrossRef]
73. Bonggotgetsakul, Y.Y.N.; Cattrall, R.W.; Kolev, S.D. The preparation of a gold nanoparticle monolayer on the surface of a polymer inclusion membrane using EDTA as the reducing agent. *J. Membr. Sci.* **2011**, *379*, 322–329. [CrossRef]
74. Lim, T.H.; McCarthy, D.; Hendy, S.C.; Stevens, K.J.; Brown, S.A.; Tilley, R.D. Real-time tem and kinetic monte carlo studies of the coalescence of decahedral gold nanoparticles. *ACS Nano* **2009**, *3*, 3809–3813. [CrossRef] [PubMed]
75. Ji, X.; Song, X.; Li, J.; Bai, Y.; Yang, W.; Peng, X. Size control of gold nanocrystals in citrate reduction: The third role of citrate. *J. Am. Chem. Soc.* **2007**, *129*, 13939–13948. [CrossRef] [PubMed]
76. Goia, D.; Matijević, E. Tailoring the particle size of monodispersed colloidal gold. *Colloids Surf. A* **1999**, *146*, 139–152. [CrossRef]
77. Weissleder, R. A clearer vision for in vivo imaging. *Nat. Biotechnol.* **2001**, *19*, 316–317. [CrossRef] [PubMed]
78. West, J.L.; Halas, N.J. Engineered nanomaterials for biophotonics applications: Improving sensing, imaging, and therapeutics. *Annu. Rev. Biomed. Eng.* **2003**, *5*, 285–292. [CrossRef] [PubMed]
79. Lee, J.W.; Jung, H.; Cho, H.H.; Lee, J.H.; Nam, Y. Gold nanostar-mediated neural activity control using plasmonic photothermal effects. *Biomaterials* **2018**, *153*, 59–69. [CrossRef] [PubMed]

nanomaterials

MDPI

Communication

Active Accumulation of Spherical Analytes on Plasmonic Hot Spots of Double-Bent Au Strip Arrays by Multiple Dip-Coating

Jinhyung Lee [1,2], Eun-Ah You [1], Do Won Hwang [3], Shinill Kang [2,*] and Jung-Sub Wi [1,*]

1 Center for Nano-Bio Measurement, Korea Research Institute of Standards and Science, Daejeon 34113, Korea; jinhyung@kriss.re.kr (J.L.); eayou@kriss.re.kr (E.-A.Y.)
2 School of Mechanical Engineering, Yonsei University, Seoul 03722, Korea
3 Department of Nuclear Medicine, Seoul National University Hospital, Seoul 03080, Korea; hdw6592@snu.ac.kr
* Correspondence: snlkang@yonsei.ac.kr (S.K.); jungsub.wi@kriss.re.kr (J.-S.W.); Tel.: +82-2-2123-2829 (S.K.); +82-42-868-5691 (J.-S.W.)

Received: 21 March 2019; Accepted: 24 April 2019; Published: 26 April 2019

Abstract: To achieve sensitive plasmonic biosensors, it is essential to develop an efficient method for concentrating analytes in hot spots, as well as to develop plasmonic nanostructures for concentrating light. In this study, target analytes were delivered to the surface of double-bent Au strip arrays by a multiple dip-coating method; they were self-aligned in the valleys between neighboring Au strips by capillary forces. As the valleys not only accommodate target analytes but also host strong electromagnetic fields due to the interaction between adjacent strips, sensitive measurement of target analytes was possible by monitoring changes in the wavelength of a localized surface plasmon resonance. Using the proposed plasmonic sensor and target delivery method, the adsorption and saturation of polystyrene beads 100 nm in size on the sensor surface were monitored by the shift of the resonance wavelength. In addition, the pH-dependent stability of exosomes accumulated on the sensor surface was successfully monitored by changing the pH from 7.4 to 4.0.

Keywords: localized surface plasmon resonance; dip-coating; capillary force; exosome

---

1. Introduction

Biosensors based on localized surface plasmon resonance (LSPR), the collective oscillation of electrons in nanostructured noble metals induced by resonant light, have been extensively investigated owing to their molecular-level sensitivity, rapid and label-free detection, and simple instrumentation [1–4]. Since LSPR sensors monitor the resonance wavelength shifts caused by analytes adsorbed on the metal surface, fabrication of uniform plasmonic nanostructures with sharp and clean resonance spectra is a prerequisite for achieving sensitive and reliable LSPR sensors [1–7]. Once well-defined plasmonic nanostructures are obtained, the delivery of analytes to plasmonic hot spots is essential to fully utilize the sensor performance. However, despite the availability of a number of reports on the fabrication of various metallic nanostructures and their application in plasmonic sensors, only a few studies have focused on the locating of analytes on sensing spots [8–10].

In this report, we present an efficient method for locating analytes in plasmonic hot spots. By a template-assisted self-assembly during the dip-coating process [11–14], target analytes spontaneously align in the valleys between neighboring Au strips, where electromagnetic fields are locally enhanced. During repeated dip-coating, analytes accumulate at the hot spots, and this phenomenon is monitored for changes in LSPR wavelength. Furthermore, we demonstrate that the proposed method can be used to monitor the stability of exosomes, extracellular vesicles (<100 nm in size), attracting much scientific and engineering interest as biomarkers for diagnosing diseases and as delivery vehicles for bioresorbable drugs [15–18].

2. Experimental Methods

2.1. Nanofabrication of Double-Bent Au Strip Arrays

The sample preparation process is illustrated in Figure 1. Initially, one-dimensional grating patterns with a period of 200 nm (1:1 line/space) and a height of 100 nm were generated on polyurethane acrylate (PUA)-coated polyethylene terephthalate (PET) substrate by UV-nanoimprint lithography. Then, a 30 nm-thick Au film was thermally evaporated at an oblique angle to the polymer nanograting. Because of the tilted deposition angle (35° from the surface normal direction) and a shadowing effect of the line grating, a double-bent Au strip (DAS) array was spontaneously formed in two steps, namely, UV-nanoimprint lithography and Au deposition. More details on the fabrication of DAS arrays can be found in our previous report, in which the refractive index sensitivity of about 210 nm refractive index unit $(RIU)^{-1}$ and a figure of merit (FOM) of 4.2 for DAS arrays were demonstrated [19].

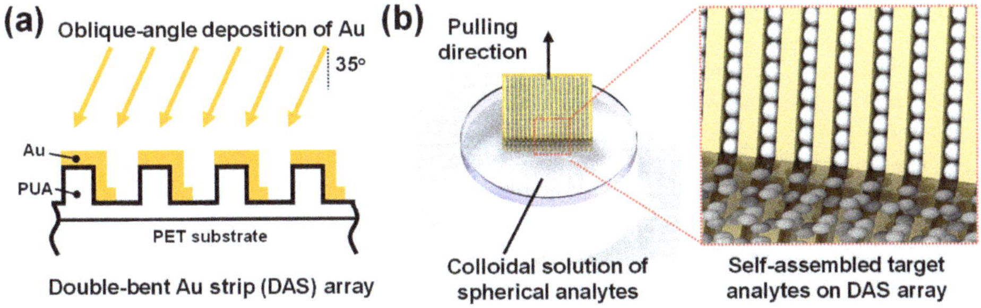

Figure 1. Schematic illustration of sample preparation. (**a**) Double-bent Au strips (DAS) were thermally deposited at an oblique angle on UV-imprinted polymer nanograting structures. PUA and PET are polyurethane acrylate and polyethylene terephthalate, respectively. (**b**) A dip-coating process was used to assemble polystyrene beads or exosomes on the DAS sensors; the dip-coating process was repeated to accumulate analytes.

2.2. Active Accumulation of Spherical Analytes on DAS Arrays by Multiple Dip-Coating

After preparation of DAS arrays, the samples were vertically dipped in a colloidal solution of polystyrene (PS) beads or exosomes and pulled out at a speed of 0.01 mm/s. A 1 wt % aqueous solution of PS beads was purchased from ThermoFisher Scientific (Nanosphere^{TM} size standard 3100A, Waltham, MA, USA). Between each coating cycle in the multiple dip-coating process, the samples were dried for 5 min in ambient atmosphere and then immersed again. After the dip-coating process was completed, the absorbance curves were measured using a UV-Vis spectrophotometer (UV-2600, Shimadzu, Kyoto, Japan) and the surfaces were observed using scanning electron microscopy (SEM, S-4800 FE-SEM, Hitachi, Tokyo, Japan).

2.3. Assessment of Exosome Stability

Exosomes were isolated from human fibroblast cells grown in Dulbecco's minimal essential medium supplemented with 10 % fetal bovine serum, 100 U/mL penicillin/streptomycin at 37 °C. The supernatant was centrifuged at 800× g for 5 min and 2000× g for 10 min to remove cellular debris. The ultracentrifugation step was conducted in cell debris-removed supernatant at 100,000× g for 3 h to collect the pellet (Beckman, Brea, CA, USA). The exosomal protein concentration of 1 μg/μL was measured using a Pierce BCA Protein Assay Kit. To examine the pH-dependent stability of exosomes, the exosome-adsorbed DAS arrays were immersed in a TRIzol reagent (Invitrogen, Waltham, MA,

USA), 1× phosphate buffer solution with pH 7.4 (Merck, Darmstadt, Germany), and citric acid/sodium hydroxide buffer solutions with pH 4.0 and 5.0 (Merck, Darmstadt, Germany).

3. Results and Discussion

3.1. Multiple Dip-Coating of Spherical Analytes on Nanoplasmonic Sensors

To examine the anisotropic wetting effect of the one-dimensional grating structure [20–22], the dipping and pulling directions of the DAS arrays were set parallel or perpendicular to the length direction of the Au strips. Figure 2 shows the SEM images of the PS bead-coated sample surfaces, depending on the direction and number of dip-coating cycles. Due to capillary forces induced on the three-phase (liquid–vapor–substrate) contact line [11–14], PS beads spontaneously assembled in the valleys between neighboring DAS structures and adsorbed on the surface by the van der Waals forces, regardless of the dip-coating direction. However, it should be noted that the density of the PS beads on the samples where the dip-coating direction was parallel to the Au strip (hereinafter referred to as par-dip sample) was much higher than that on the samples with a perpendicular dip-coating direction (hereinafter referred to as the per-dip sample). For example, in the case of a triple dip-coating, the population of PS beads of the par-dip sample was 154 in the SEM image (Figure 2f), which was approximately five times larger than that of the per-dip sample (Figure 2b). It is believed that this difference was caused by anisotropic wettability and evaporation-induced flow near the contact line. Because the liquid contact angle in the direction parallel to the Au strip was smaller than the contact angle along the perpendicular direction, the par-dip sample with its long meniscus could secure a longer time for positioning the PS beads than the per-dip sample. In addition, the contact line with its long meniscus could induce an outward flow of the PS beads during evaporation; this phenomenon is known as the coffee stain phenomenon [23,24]. Accordingly, deposits of self-assembled beads could accumulate much more on the par-dip sample than on the per-dip sample during the multiple dip-coating process. In the case of the par-dip sample, the valleys between neighboring DAS structures were almost filled with PS beads after five repeated dip-coating cycles, as shown in Figure 2g.

The valleys between the Au strips not only accommodate PS beads by capillary forces, but also act as hosts to strong plasmonic fields due to the interactions between adjacent strips, as was demonstrated in experimental and simulation studies [25]. Therefore, the PS beads self-assembled in the valley between the DAS arrays can induce a change in the resonance conditions of the Au strips, resulting in a spectral shift of the LSPR wavelength. Figure 3 shows the LSPR shifts of the per- and par-dip samples depending on the number of dip-coating repetitions. The LSPR peak shifts of the par-dip samples were much larger than those of the per-dip samples due to a greater number of dense bead deposits on the sample surfaces. This observation is in good agreement with the SEM images in Figure 2. As the LSPR peak shift was accompanied by a change in the amplitude of absorbance, as shown in Figure 3a,b, it would be possible to monitor the change in absorbance at a specific wavelength (e.g., the initial resonant wavelength) as an alternative signal reading method.

Figure 3a,c also indicates that the peak shift generated by the dip-coating was reduced after five repetitions of the dip-coating process. Once the valleys were filled with PS beads, the additional beads adsorbed on the upper sides of the DAS structures were too far away from the plasmonic fields developed in the valleys, and hence they did not significantly affect the resonant condition of the Au strips. For the same reason, analytes that are too small or beads that are too large compared to the dimensions of the valley are not very sensitive to DAS arrays, as demonstrated by experiments and theoretical calculations in our previous report [25]. Therefore, the dimensions of the DAS structures can be adapted to the size of a particular analyte, and the gold surface of DAS can be functionalized with antibodies for target-specific detection.

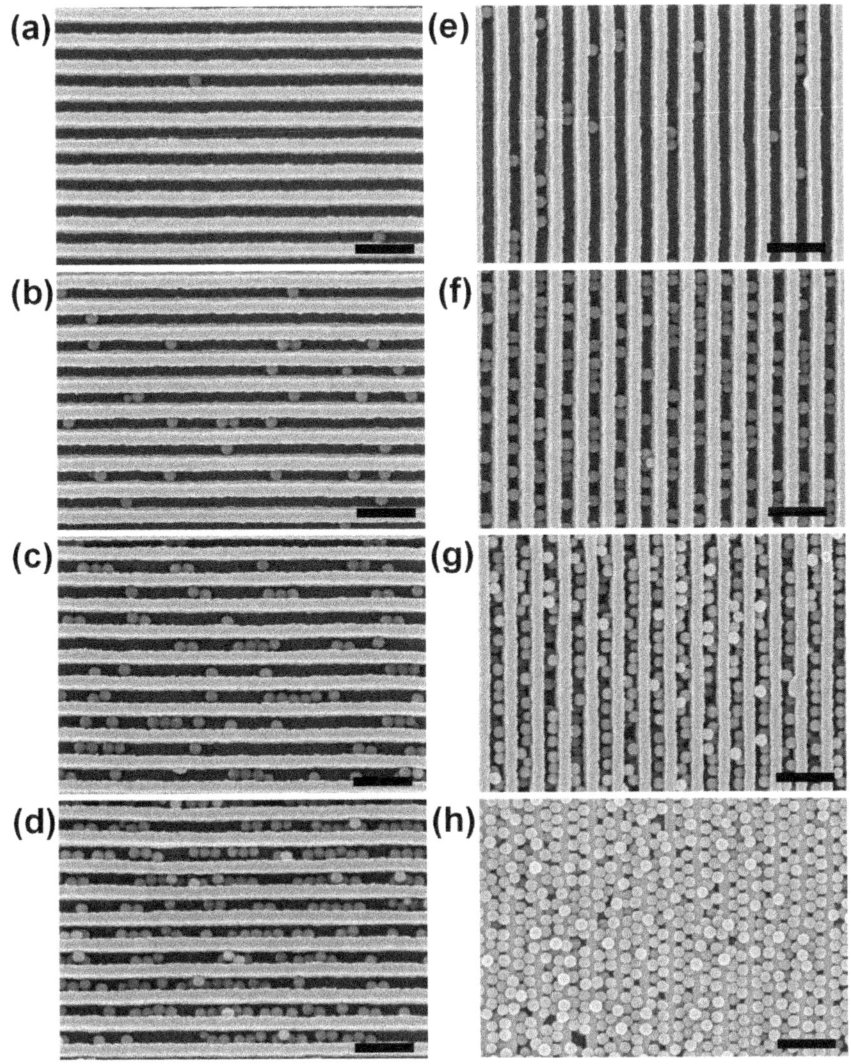

Figure 2. SEM images of the polystyrene (PS) bead-coated sample surfaces according to the direction and number of repeated dip-coating cycles. The dipping and pulling directions of DAS arrays were (**a**–**d**) perpendicular and (**e**–**h**) parallel to the direction of the length of Au strips. Samples were dip-coated (**a**,**e**) once, (**b**,**f**) three times, (**c**,**g**) five times, and (**d**,**h**) seven times. Scale bars represent 500 nm.

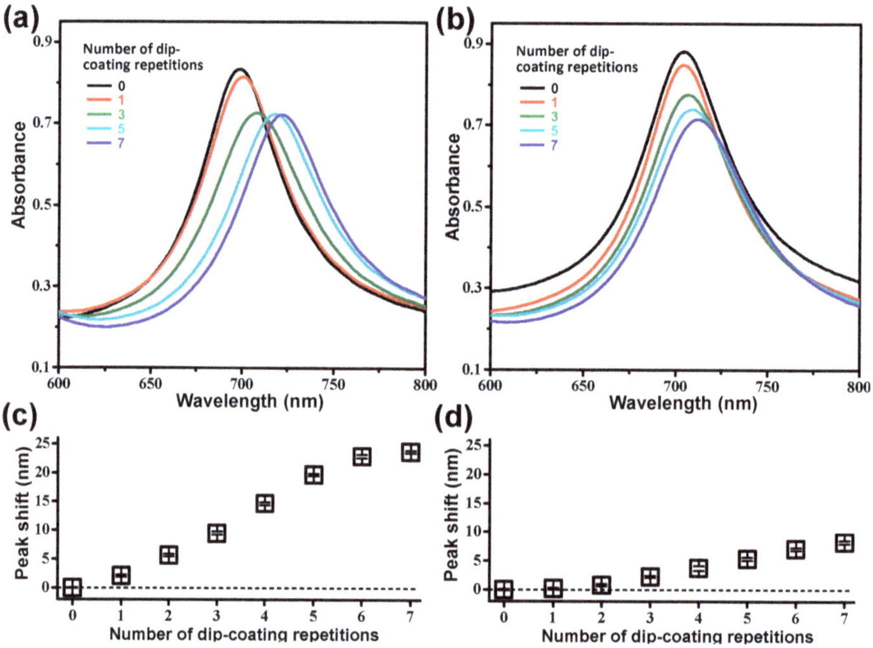

Figure 3. (**a**,**b**) Absorbance curves and (**c**,**d**) localized surface plasmon resonance (LSPR) peak shifts of PS bead-coated DAS arrays as functions of the number of dip-coating repetition cycles. The data in Figure 3a,c and Figure 3b,d were obtained from the par-dip and per-dip samples, respectively. All the data in Figure 3c,d showed statistically significant differences (P-value < 0.05), except for the first two data in Figure 3d (single and double dip coatings).

3.2. Assessment of Exosome Stability by Nanoplasmonic Sensors

In addition to monitoring the adsorption of analytes, it is also possible to monitor desorption or degradation of analytes adsorbed on DAS arrays by measuring the LSPR shifts. Thus, DAS arrays in combination with the proposed multiple dip-coating method were used to assess the stability or degradability of exosomes in neutral or acidic microenvironment known to be associated with their drug delivery efficacy [17,18]. Exosomes dispersed in a phosphate-buffered solution at a concentration of 1 μg/μL were accumulated on DAS arrays by thrice dip-coating. The dip-coated exosomes resulted in an 8 nm red shift of the LSPR peak (Supplementary Materials, Figure S1). For the case when a 10 μL drop of the same exosome sample was dried on a DAS array, the LSPR peak shift was less than 2 nm (Supplementary Materials, Figure S2a).

To examine the pH-dependent stability of exosomes, the exosome-adsorbed DAS arrays were immersed in a TRIzol reagent and buffer solutions with pH 7.4, 5.0, and 4.0. TRIzol reagent for complete fragmenting of exosomes was used as a control [26]. After treatment with TRIzol reagent, the LSPR peak of the exosome-adsorbed DAS array was blue-shifted (9.5 nm) and, thus, returned to its original position within 1 h (red bars in Figure 4a). The 1.5 nm blue shift from the original position may be the result of damage to the polymer substrate (PET) by the reagent. Compared with TRIzol reagent, it was confirmed that buffer solutions with pH 7.4, 5.0, and 4.0 did not make any noticeable damage on the DAS arrays and the substrate (Supplementary Materials, Figure S2b). In the buffer solution of pH 7.4, the physiological condition was stable for exosomes, and no apparent changes in the LSPR peak were observed, although the exosome-coated DAS array was immersed for up to 60 h (black bars in Figure 4a). Compared to the previous two extremely unstable (TRIzol) and stable (pH 7.4) cases, the peak shifts of the samples immersed in acidic buffer solutions (pH 4.0 and 5.0) exhibited gradual changes, as shown in

Figure 4b. The gradual blue shift in the LSPR peaks indicates that the exosomes on the DAS array were partially desorbed from the sample surface or degraded under acidic conditions with an increase in the immersion time. As the desorption of exosomes in the pH 7.4 buffer solution was not very significant, exosome degradation is considered a possible cause of the blue shift of the LSPR peak in acidic buffer solutions. Petelsak et al. demonstrated that a lipid bilayer membrane is subject to greater interfacial tension under acidic conditions than under neutral conditions [27]. Therefore, in pH 4.0 and pH 5.0 buffer solutions, increased interfacial tension of the lipid membrane can reduce the stability of exosomes and result in a blue shift of the LSPR peak. In addition to testing the exosome stability demonstrated in this report, it is possible to monitor the loading and unloading of drug molecules from the exosomes under various physiological conditions that are the subjects of our further studies.

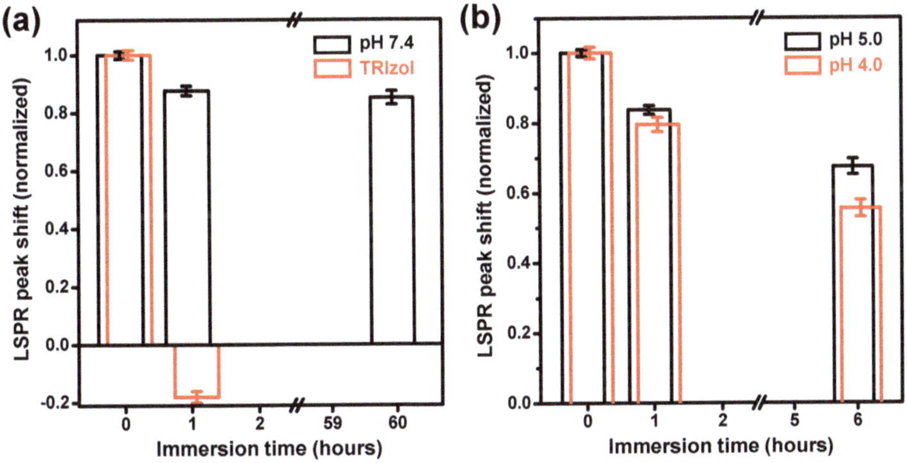

Figure 4. Normalized LSPR peak shifts measured by exosome-coated DAS arrays. After dip-coating thrice with exosomes, the samples were immersed in a TRIzol reagent (red bars in (**a**)) and buffer solutions with pH 7.4 (black bars in (**a**)), 5.0 (black bars in (**b**)), and 4.0 (red bars in (**b**)). The peak shifts were divided by the initial shift values, measured after dip-coating thrice with exosomes for each sample.

4. Conclusions

For effective placement of target analytes on sensor surfaces and their monitoring, we present a straightforward measurement method based on nanoplasmonic sensors of double-bent Au strip arrays, multiple dip-coating of analytes, self-alignment of analytes in the region of strong plasmonic fields, and spectrometric monitoring of LSPR peaks. Using this method, closely packed PS beads in the valleys of the DAS array and pH-dependent stability of the exosomes were successfully monitored in terms of shifts in the LSPR peaks. As a small amount of target analytes can accumulate in the plasmonic hot spot due to multiple dip-coating cycles, and the LSPR peak can be measured with a conventional UV-vis spectrometer under physiological conditions, it is expected that the proposed measurement platform will be useful for studying the stability of various drug delivery vesicles and their efficiencies.

Supplementary Materials: The following are available online at http://www.mdpi.com/2079-4991/9/5/660/s1, Figure S1: (a) Measured absorbance curves of the DAS array (black) before and (red) after dip-coating thrice with exosomes and (blue) after immersion in a TRIzol reagent for 1 h. (b) Magnified view of Figure S1a. Figure S2: (a) Measured absorbance curves of the DAS array (black) before and (red) after drying a 10 μL drop of an aqueous solution of exosome. (b) Table of LSPR peak shifts of the DAS arrays, measured after dipping in pH buffer solutions with pH 7.4, 5.0, and 4.0 for 18 hours.

Author Contributions: Conceptualization, J.-S.W.; methodology, J.-S.W.; validation, J.L. and J.-S.W.; formal analysis, J.L. and J.-S.W.; investigation, J.L. and J.-S.W.; resources, E.-A.Y. and D.W.H.; data curation, J.L. and

J.-S.W.; writing—original draft preparation, J.-S.W.; writing—review and editing, S.K. and J.-S.W.; visualization, J.L.; supervision, S.K.; project administration, S.K. and J.-S.W.; funding acquisition, J.-S.W.

Funding: This study was supported by the Korea Research Institute of Standards and Science (Development of Platform Technology for Innovative Medical Measurements [KRISS-2019-GP2019-0013]) and by the National Research Foundation funded by the Ministry of Science and ICT of Korea (Bio and Medical Technology Development Program [2015M3A9D7029894], Global Frontier Project [HGUARD_2013M3A6B2078962], and Nano Material Technology Development Program [2014M3A7B6020163, 2017M3A7B4041754]).

Conflicts of Interest: The authors declare no conflict of interest.

References

1. Willets, K.A.; Van Duyne, R.P. Localized Surface Plasmon Resonance Spectroscopy and Sensing. *Annu. Rev. Phys. Chem.* **2007**, *58*, 267–297. [CrossRef] [PubMed]
2. Stewart, M.E.; Anderton, C.R.; Thompson, L.B.; Maria, J.; Gray, S.K.; Rogers, J.A.; Nuzzo, R.G. Nanostructured plasmonic sensors. *Chem. Rev.* **2008**, *108*, 494–521. [CrossRef] [PubMed]
3. Estevez, M.C.; Otte, M.A.; Sepulveda, B.; Lechuga, L.M. Trends and challenges of refractometric nanoplasmonic biosensors: A review. *Anal. Chim. Acta* **2014**, *806*, 55–73. [CrossRef] [PubMed]
4. Špačková, B.; Wrobel, P.; Bocková, M.; Homola, J. Optical Biosensors Based on Plasmonic Nanostructures: A Review. *Proc. IEEE* **2016**, *104*, 2380–2408. [CrossRef]
5. Shen, Y.; Zhou, J.; Liu, T.; Tao, Y.; Jiang, R.; Liu, M.; Xiao, G.; Zhu, J.; Zhou, Z.-K.; Wang, X.; et al. Plasmonic Gold Mushroom Arrays with Refractive Index Sensing Figures of Merit Approaching the Theoretical Limit. *Nat. Commun.* **2013**, *4*, 2381. [CrossRef] [PubMed]
6. Wi, J.-S.; Tominaka, S.; Nagao, T. Arrays of Nanoscale Gold Dishes Containing Engineered Substructures. *Adv. Opt. Mater.* **2013**, *1*, 814–818. [CrossRef]
7. Sreekanth, K.V.; Alapan, Y.; ElKabbash, M.; Ilker, E.; Hinczewski, M.; Gurkan, U.A.; De Luca, A.; Strangi, G. Extreme Sensitivity Biosensing Platform Based on Hyperbolic Metamaterials. *Nat. Mater.* **2016**, *15*, 621–627. [CrossRef]
8. Le Ru, E.C.; Grand, J.; Sow, I.; Somerville, W.R.C.; Etchegoin, P.G.; Treguer-Delapierre, M.; Charron, G.; Félidj, N.; Lévi, G.; Aubard, J. A scheme for detecting every single target molecule with surface-enhanced raman spectroscopy. *Nano Lett.* **2011**, *11*, 5013–5019. [CrossRef]
9. Huang, M.; Galarreta, B.C.; Cetin, A.E.; Altug, H. Actively transporting virus like analytes with optofluidics for rapid and ultrasensitive biodetection. *Lab. Chip* **2013**, *13*, 4841–4847. [CrossRef]
10. Wang, W.; Yin, Y.; Tan, Z.; Liu, J. Coffee-ring effect-based simultaneous SERS substrate fabrication and analyte enrichment for trace analysis. *Nanoscale* **2014**, *6*, 9588–9593. [CrossRef]
11. Xia, Y.; Yin, Y.; Lu, Y.; McLellan, J. Template-Assisted Self-Assembly of Spherical Colloids into Complex and Controllable Structures. *Adv. Funct. Mater.* **2003**, *13*, 907–918. [CrossRef]
12. Alivisatos, A.P.; Boussert, B.; Liddle, J.A.; Cui, Y.; Sönnichsen, C.; Björk, M.T. Integration of Colloidal Nanocrystals into Lithographically Patterned Devices. *Nano Lett.* **2004**, *4*, 1093–1098. [CrossRef]
13. Grzelczak, M.; Vermant, J.; Furst, E.M.; Liz-Marzán, L.M. Directed self-assembly of nanoparticles. *ACS Nano* **2010**, *4*, 3591–3605. [CrossRef]
14. Choe, J.H.; Park, Q.H.; You, E.A. Rational and Facile Construction of 3D Annular Nanostructures with Tunable Layers by Exploiting the Diffraction and Interference of Light. *Adv. Funct. Mater.* **2016**, *26*, 5203–5210. [CrossRef]
15. Théry, C.; Zitvogel, L.; Amigorena, S. Exosomes: Composition, biogenesis and function. *Nat. Rev. Immunol.* **2002**, *2*, 569–579. [CrossRef]
16. Johnstone, R.M. Exosomes biological significance: A concise review. *Blood Cells Mol. Dis.* **2006**, *36*, 315–321. [CrossRef]
17. Vader, P.; Mol, E.A.; Pasterkamp, G.; Schiffelers, R.M. Extracellular vesicles for drug delivery. *Adv. Drug Deliv. Rev.* **2016**, *106*, 148–156. [CrossRef]
18. Luan, X.; Sansanaphongpricha, K.; Myers, I.; Chen, H.; Yuan, H.; Sun, D. Engineering exosomes as refined biological nanoplatforms for drug delivery. *Acta Pharmacol. Sin.* **2017**, *38*, 754–763. [CrossRef]
19. Wi, J.S.; Lee, S.; Lee, S.H.; Oh, D.K.; Lee, K.T.; Park, I.; Kwak, M.K.; Ok, J.G. Facile three-dimensional nanoarchitecturing of double-bent gold strips on roll-to-roll nanoimprinted transparent nanogratings for flexible and scalable plasmonic sensors. *Nanoscale* **2017**, *9*, 1398–1402. [CrossRef]

20. Morita, M.; Koga, T.; Otsuka, H.; Takahara, A. Macroscopic-wetting anisotropy on the line-patterned surface of fluoroalkylsilane monolayers. *Langmuir* **2005**, *21*, 911–918. [CrossRef]
21. Sommers, A.D.; Jacobi, A.M. Creating micro-scale surface topology to achieve anisotropic wettability on an aluminum surface. *J. Micromech. Microeng.* **2006**, *16*, 1571–1578. [CrossRef]
22. Xia, D.; Brueck, S.R.J. Strongly Anisotropic Wetting on Surfaces. *Nano Lett.* **2008**, *8*, 2819–2824. [CrossRef]
23. Deegan, R.D.; Bakajin, O.; Dupont, T.F.; Huber, G.; Nagel, S.R.; Witten, T.A. Capillary flow as the cause of ring stains from dried liquid drops. *Nature* **1997**, *389*, 827–829. [CrossRef]
24. Li, Y.; Yang, Q.; Li, M.; Song, Y. Rate-dependent interface capture beyond the coffee-ring effect. *Sci. Rep.* **2016**, *6*, 1–8. [CrossRef]
25. Wi, J.-S.; Oh, D.K.; Kwak, M.K.; Ok, J.G. Size-dependent detection sensitivity of spherical particles sitting on a double-bent gold strip array. *Opt. Mater. Express* **2018**, *8*, 1774. [CrossRef]
26. Eldh, M.; Lötvall, J.; Malmhäll, C.; Ekström, K. Importance of RNA isolation methods for analysis of exosomal RNA: Evaluation of different methods. *Mol. Immunol.* **2012**, *50*, 278–286. [CrossRef]
27. Petelska, A.D.; Figaszewski, Z.A. Effect of pH on the interfacial tension of lipid bilayer membrane. *Biophys. J.* **2000**, *78*, 812–817. [CrossRef]

nanomaterials

MDPI

Review

Biomedical Applications of Silver Nanoparticles: An Up-to-Date Overview

Alexandra-Cristina Burduşel [1], Oana Gherasim [2,3], Alexandru Mihai Grumezescu [2], Laurenţiu Mogoantă [4], Anton Ficai [2] and Ecaterina Andronescu [2,*]

1 Faculty of Engineering in Foreign Languages, University Politehnica of Bucharest, 313 Splaiul Independenţei, Bucharest 060042, Romania; alexandra_burdu@yahoo.com.sg

2 Department of Science and Engineering of Oxide Materials and Nanomaterials, Faculty of Applied Chemistry and Materials Science, University Politehnica of Bucharest, 1-7 Gheorghe Polizu Street, Bucharest 011061, Romania; oana.fufa@gmail.com (O.G.); grumezescu@yahoo.com (A.M.G.); anton.ficai@upb.ro (A.F.)

3 Lasers Department, National Institute for Lasers, Plasma and Radiation Physics, 409 Atomiştilor Street, Magurele 077125, Romania

4 Research Center for Microscopic Morphology and Immunology, University of Medicine and Pharmacy of Craiova, 2 Petru Rareş Street, Craiova 200349, Romania; laurentiu_mogoanta@yahoo.com

* Correspondence: ecaterina.andronescu@upb.ro; Tel.: +40-21-318-1000

Received: 1 August 2018; Accepted: 28 August 2018; Published: 31 August 2018

Abstract: During the past few years, silver nanoparticles (AgNPs) became one of the most investigated and explored nanotechnology-derived nanostructures, given the fact that nanosilver-based materials proved to have interesting, challenging, and promising characteristics suitable for various biomedical applications. Among modern biomedical potential of AgNPs, tremendous interest is oriented toward the therapeutically enhanced personalized healthcare practice. AgNPs proved to have genuine features and impressive potential for the development of novel antimicrobial agents, drug-delivery formulations, detection and diagnosis platforms, biomaterial and medical device coatings, tissue restoration and regeneration materials, complex healthcare condition strategies, and performance-enhanced therapeutic alternatives. Given the impressive biomedical-related potential applications of AgNPs, impressive efforts were undertaken on understanding the intricate mechanisms of their biological interactions and possible toxic effects. Within this review, we focused on the latest data regarding the biomedical use of AgNP-based nanostructures, including aspects related to their potential toxicity, unique physiochemical properties, and biofunctional behaviors, discussing herein the intrinsic anti-inflammatory, antibacterial, antiviral, and antifungal activities of silver-based nanostructures.

Keywords: silver nanoparticles; biomedical applications; biological interactions; biofunctional performances; intrinsic anti-inflammatory activity; antimicrobial efficiency

1. Introduction

In the past few decades, tremendous interest and substantial research efforts were directed toward the biomedical evaluation and revaluation of metallic nanoparticles derived from noble metals, such as silver and gold, thanks to their specific and genuine chemical, biological, and physical properties [1,2]. In particular, impressive attention was oriented toward the biomedicine-related assessment of silver nanoparticles (AgNPs), which first attracted worldwide attention as unconventional antimicrobial agents [3–5]. Even though there is limited information regarding the toxicity and in vivo biological behavior of AgNPs, these nanostructures were used for a long time as antibacterial agents in the health industry [6,7], cosmetics [8,9], food storage [10,11], textile coatings [12,13], and some environmental applications [14–16]. AgNPs are a class of zero-dimensional materials with distinctive morphologies, having a size ranging from 1 nm to 100 nm [17].

As to the methods of obtaining AgNPs, different strategies were successfully used, thanks to the intrinsic versatility of silver metal and silver-based compounds, including physical [18,19], chemical [20,21], physicochemical [22,23], and biological synthesis approaches [24,25]. However, given the facile and safe process, reduced economic implications, and repeatability and reproducibility of experimental results, the method most used in the preparation of AgNPs is represented by the chemical reduction of silver salts by sodium citrate or sodium borohydrate [26,27]. In addition to their intrinsic antimicrobial-related applications, AgNPs were thoroughly explored thanks to their beneficial size-related physicochemical effects exhibited in novel electronic, magnetic, catalytic, and optical materials [28,29].

Special interest is oriented toward improving the stability of AgNPs, since a particular limitation of their antimicrobial-related use arises from their instability in bacteria-rich environments, and consequently, diminution or deprivation of their anti-pathogenic activity. In order to improve the stability of AgNPs in solution, many inorganic and organic [30,31], synthetic and natural [32,33], and biotic and abiotic materials were used as capping agents [34].

Though the precise anti-pathogenic mechanism of silver nanoparticles remains to be clarified, it is postulated that nanosilver-based systems exert their antimicrobial effects through the following phenomena: (a) microbial membrane damage, caused by the physicochemically guided attachment of AgNPs on the cell surface, and subsequent structural and functional alterations (such as gap formation, membrane destabilization, membrane piercing, and cytoplasm leakage); and (b) microbial sub-cellular structure damage, caused by the release of free Ag^+ ions and subsequent reactive oxygen species (ROS) generation or essential macromolecule (proteins, enzymes, and nucleotides) inactivation [35–37]. Still, the most remarkable mechanistic mode of AgNP-based antimicrobial effects is represented by their adhesion to microbial cells, ROS and free-radical generation, microbial wall piercing and penetration inside cells, and modulation and modification of microbial signal-transduction pathways [38]. Metallic silver ions are strong antimicrobials themselves, but they are easily isolated by phosphate and chloride functions, proteins, and different cellular components [39]. The intrinsic biocide or biostatic activity of AgNPs is strongly influenced by different physicochemical features, including morphology, size, oxidation and dissolution states, surface charge, and surface coating [37,40].

The effectiveness of nanosilver-based biomaterials as promising antimicrobial agents was experimentally assessed against a wide range of medically relevant planktonic and sessile pathogenic microorganisms, including bacteria [41,42], viruses [43,44], fungi, and yeasts [45,46]. The impressive antimicrobial activity of AgNPs is a solid starting point for the design, development, and implementation of new and performance-enhanced nanosilver-based biomedical products, such as anticancer agents, drug-delivery platforms, orthopedic materials and devices [47], bandages, antiseptic sprays, and catheters [48]. As a consequence of the impressive applicability of AgNPs in the fields of nanotechnology, biomedicine, and environment, there is a continuous need for the development of cost-effective methods for the synthesis of AgNPs [49]. The translation of silver-based nanotechnology to clinical applications requires not only the development of safe, simple, eco-friendly, and cost-effective methods for the synthesis of silver nanoparticles, but also a thorough understanding of the related physicochemical particularities, in vitro and in vivo effects, biodistribution, safety control mechanisms, pharmacokinetics, and pharmacodynamics of AgNPs [48].

2. Antibacterial Characteristics of Silver Nanoparticles

Silver nanoparticles attracted tremendous interest in the biomedical field, thanks to their attractive and unique nano-related properties, including their high intrinsic antimicrobial efficiency and non-toxic nature. Among the manifold potential applications of AgNPs in this particular domain, impressive attention and efforts were lately directed toward their promising implications in wound dressing, tissue scaffold, and protective clothing applications [50,51]. Some essential aspects related to the specific antimicrobial characteristics of AgNPs implies their intrinsic physical and chemical properties, which include maintaining the nanoscale size of AgNPs, improving their dispersion and stability, and avoiding aggregation [52]. There are many studies which experimentally proved that the anti-pathogenic activity of AgNPs is better than that exhibited by silver ions [53].

A major concern of the worldwide healthcare system is represented by the alarming and emerging phenomenon of pathogenic drug-resistant occurrence. Therefore, AgNPs represent potent candidates for the nanotechnology-derived development of novel and effective biocompatible nanostructured materials for unconventional antimicrobial applications [54]. Thanks to their intrinsic broad bactericidal effects exhibited against both Gram-negative and Gram-positive bacteria and their physicochemical properties, AgNPs are one of the most used metallic nanoparticles in modern antimicrobial applications [55]. Different studies reported that AgNPs interact with the bacterial membrane and penetrate the cell, thus producing a drastic disturbance regarding proper cell function, structural damage, and cell death [56]. We included in Figure 1 distinctive mechanisms described during the interaction of AgNPs with bacterial cells [57].

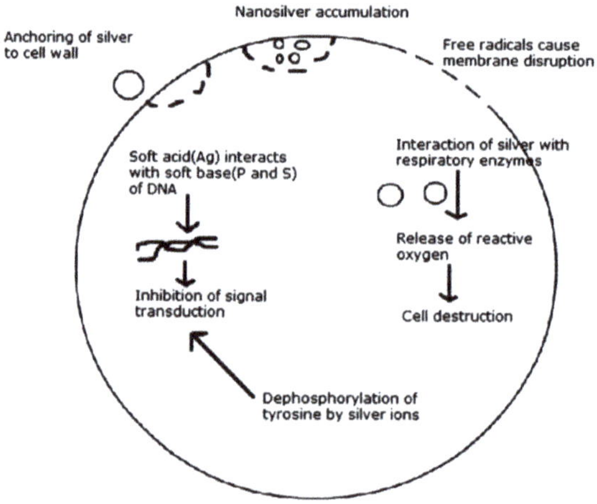

Figure 1. Various modes of action of silver nanoparticles (AgNPs) on bacteria [57].

Currently, it is shown that AgNPs can be successfully used to design and develop improved wound and burn dressings, thanks to the intrinsic antibacterial and anti-inflammatory effects of sole metallic nanoparticles [58]. Given the mutation-resistant antimicrobial activity related to nanosilver-based biomaterials, AgNPs are used in various pharmaceutical formulations as burn ointments, antibacterial clothing, and coatings for medical devices [59]. Different studies proved that the stabilization of AgNPs against dissolution and/or agglomeration can be achieved by using various capping agents, such as sodium citrate, polyvinylpyrrolidone (PVP), or polyethylene glycol (PEG) [60]. The stability of AgNPs was previously investigated, where the reported data indicated that their stability actively influences their toxicity [60].

The most important physicochemical parameters that affect the antimicrobial effects exhibited by AgNPs include size [61], shape [62], concentration [63], surface charge [64], and colloidal state [65]. As mentioned before, AgNPs display their intrinsic enhanced antimicrobial activity through various mechanisms [66].

It is also worth mentioning that the AgNP-based treatment of human cell cultures may induce cytotoxicity [67], inflammatory responses in a cell-type-dependent manner, and genotoxicity [66]. Thanks to their intrinsic capacity to provide stimuli-dependent responses through the specific modification of their optical properties, chemical environments, high molar absorptivity [68], and the many sorption sites found on their extensive surface [69], AgNPs are also used in various analytical applications. There are manifold research studies which particularly describe the biocide activity of silver itself [70].

The current state of the art relies on the beneficial conjunction between antimicrobial silver nanoparticles and natural or synthetic polymers, in the modern attempt to diminish or even eliminate the microbial contamination and colonization processes [71].

The major advantage of nanosilver-based biomaterials designed for unconventional antibacterial applications is related to their intrinsic anti-pathogenic effects exhibited against both planktonic and biofilm-organized microorganisms. The bactericidal activity of AgNPs is attributed to silver cations, which possess the ability to specifically bind to thiol groups of bacterial proteins, disrupting their physiological activity and leading to cell death. The effects of AgNPs on bacterial DNA were not analyzed in detail with respect to possible DNA lesions and antibacterial action that occur after AgNPs treatment. Silver nanoparticles exert their bactericidal activity through a Trojan-horse mechanism, since their initial binding to the cell surface leads to permeability alteration and cellular respiration impairment, followed by cell-barrier penetration and intracellular metallic silver ion release.

In order to successfully apply nanosilver-based systems as effective antibacterial agents, it is important to thoroughly understand their action against bacterial cells and bacterial biofilms [72]. In addition to their improved efficiency against planktonic bacteria (as discussed above), AgNPs also possess bactericide or bacteriostatic activity against biofilm-organized microorganisms. The antibacterial effects exhibited by silver-based nanosystems against biofilm-organized bacteria may be due to intrinsic activity against isolated or block cells, destabilization or disruption of the exopolymeric substances within the extracellular biofilm matrix, or interfering with bacterial signaling molecules [73–75]. There are ongoing discussions regarding the role of Ag ions released from AgNPs and their related toxic effect on microorganisms. Many researchers stated that the toxicity of AgNPs is due to the nanoparticles themselves, whereas others provided evidence that silver ions released from AgNPs play a crucial role during antimicrobial activity. Following the release of silver ions from AgNPs, antibacterial activity is initiated by metallic cations, rather than by metallic nanoparticles [76].

3. Silver Nanoparticles for Drug-Delivery Systems

In medicine, the pharmacokinetics and pharmacodynamics of drugs are as important as their intrinsic therapeutic effects [77]. Since the specific and selective delivery and action of therapeutic agents became one of the most studied topics for improving current human healthcare practice, nanoparticles received tremendous attention regarding the design and development of novel and enhanced drug-delivery systems [78]. In particular, AgNP-based nanosystems were evaluated as suitable carriers of various therapeutic molecules, including anti-inflammatory [79,80], anti-oxidant [81,82], antimicrobial [83,84], and anticancer [32,85] biosubstances.

In order to provide a specific therapeutic effect in human or animal organisms, it is essential to consider the process or method applied during the administration of the selected pharmaceutical compound [86]. For obtaining novel and performance-enhanced drug-delivery systems responsive to thermal, optical, or pH modulations to target inflammatory, infectious, and malignant ailments, hybrid molecular units consisting of AgNPs were successfully chosen, especially thanks to their exceptional biocompatibility and viable features for nanoscale-derived therapeutic settings [87].

As a consequence of the difficulty encountered during AgNP synthesis and the concerns regarding the toxicity and reduced stability of nanosilver-based systems when functionalized according to conventional salt-aging techniques, silver is not extensively used in nanoparticle-based drug-delivery applications, instead being replaced with gold or other nanomaterials [88]. An excellent triggerable and tunable nanosystem for drug-delivery applications should be easy to develop from readily available components, exhibit optimal responsiveness, and be compatible with more than one trigger [89]. Moreover, such a particular drug-delivery [90] platform should not only provide suitable and adjustable drug loading and releasing profiles, but should also enable [70] maximal therapeutic efficiency at concentrations below that of the sole biosubstance with side-effect minimization [91,92].

Thanks to their intrinsic anticancer activity [93], AgNPs attracted special attention for this particular domain, and were successfully evaluated as effective anti-tumor drug-delivery systems [94], acting either as passive [95,96] or active [97,98] nanocarriers for anticancer drugs. For the preparation of biocompatible AgNPs, different strategies were used, such as organic-water two-phase synthesis [99–101], micro-emulsion [102–104], radiolysis [105,106], and most commonly, reduction in aqueous solution [94,107,108]. Impressive attention, scientific knowledge, and financial support were lately oriented toward the formulation of AgNP-based drug-delivery platforms, thanks to the intrinsic features of nanosilver, including its capacity to bind a wide range of organic molecules, its tunable and strong absorption properties, and its low toxicity [109]. Recent studies evidenced the potential use of AgNPs as vaccine and drug carriers for specific and selective cell or tissue targeting [109]. In addition to the great optical properties of AgNPs (governed by specific surface plasmon resonance and localized surface plasmon resonance) [110–112], the recent improvements in AgNP biocompatibility and stability via surface modification strongly recommend nanostructured systems based on silver as specific, selective, and versatile candidates for drug-delivery applications [113].

4. Silver Nanoparticles for Catheter Modification

Central venous catheters (CVC) were firstly described by Niederhuber in 1982; since then, these devices became important therapeutic tools for diverse clinical conditions requiring malnutrition and replacement therapy (e.g., renal disease and cancer) [114]. CVCs are normally used to provide access for intravenous fluid administration, hemodynamics monitoring, drug-delivery pathways [115], and nutritional support in critically ill patients. Still, these medical devices are also a considerable source of hospital-acquired infections [116], and are considered a specific high-risk category of devices susceptible to microbial contamination and colonization phenomena [117]. A recent study showed that various *Staphhylococcus aureus* strains are responsible for catheter-related infections, and 82% of them are methicillin-resistant strains possessing many genes expressed in biofilm development and bacterial dispersion processes [118].

In order to induce antibacterial effects to clinically relevant materials and devices, AgNPs were extensively explored for the modification of one-dimensional and two-dimensional surfaces [119], such as cotton fabrics [120,121], natural and artificial fibers [122–124], thin polymer films [125,126], and wound pads [127,128].

Even if silver (a half-noble metal) is susceptible to quick oxidation processes, the impressive surface-to-volume atomic ratio related to AgNPs accounts for the sustained local supply of Ag^+ ions at the coating/tissue interface [129]. In recent studies, the role of AgNP-modified catheters as non-toxic devices capable of sustained release of bactericidal silver, exhibiting preventive effects against infection-related complications, was presented [116,130,131]. Given the fact that one of the major groups of organisms that causes device-related infections is represented by coagulase-negative staphylococci (CoNS), the effects exhibited by AgNPs and AgNP-coated catheters against these organisms were intimately studied [38]. Significant inhibitory effects against both Gram-positive and Gram-negative bacterial biofilm development were exhibited by CVCs coated with AgNPs [115,132–134].

Because the binding capacity of silver nanoparticles to bacterial cells is influenced by the surface area available for interaction, the bactericidal effects are expected to be size-dependent [135]. Catheters treated with silver ions represent a feasible strategy for reducing dialysis-related infections in patients undergoing peritoneal catheters; however, the antimicrobial efficiency and obtaining methods of Ag^+ are different [136]. Silver/copper-coated catheters were assessed as a promising solution for preventing methicillin-resistant *Staphylococcus aureus* (MRSA) infections, since their antibacterial activity might be improved by limiting non-specific plasma protein adsorption [137].

The main complication related to urinary catheterization is represented by the occurrence of catheter-associated urinary tract infections (CAUTIs) [138]. It was shown that a polymer matrix impregnated with AgNPs displayed hydrophilic surface properties, resulting in the prevention of bacterial biofilm formation and the deposition of proteins and electrolytes responsible for incrustation and adherence of microorganisms onto the surface [139]. With regards to silicon urethral catheters, Kocuran-capped silver glyconanoparticles were successfully evaluated as effective antibiofilm and antimicrobial coatings [118]. Despite the concerns regarding CVC-related complacency with respect to septic techniques, catheters with antimicrobial properties were taken into consideration as a feasible means of supplying additional protection against microbial contamination, further reducing colonization and infection risks [117].

5. Silver Nanoparticles for Dental Applications

Dental caries represent one of the most extensive oral-cavity-related affections worldwide, being also an economic burden [140]. By enhancing the remineralization process and controlling biofilm development, nanotechnology-derived dental-related strategies aim to limit or even eliminate the clinical impact of caries [140]. In addition to their intrinsic highly biocompatible behavior, the materials for dental barrier membranes (DBM), which are often used for efficient alveolar bone reconstruction, must accomplish some specific and additional features and functions [141]. Different metal-coated implants were evaluated against various pathogens responsible for dental-related biofilm formation and subsequent implant failure [142].

In order to prevent the pathogenic contamination of dental implants, proper tooth-brushing techniques, prophylactic antibiotics, and antimicrobial mouthwashes are specifically recommended [143]. A major goal in dentistry is to provide the proper protection of the oral cavity, which represents a pathogenic-susceptible gateway for the entire body [144]. Biofilms developed on dental implant surfaces may additionally cause inflammatory lesions on the peri-implant mucosa, thus increasing the risk of implant failure [145].

Silver was used for centuries in oral care and gained worldwide attention in the 19th century, being a major component in dental amalgams used for tooth restoration [146]. AgNPs were also used in various fields of dentistry, such as dental prostheses, restorative and endodontic dentistry, and implantology [147]. Thanks to their unique properties feasible for different domains of real interest in modern society, silver nanoparticles hold a prominent place in nanomaterial-related restorative, regenerative, and multifunctional biomedicine [148,149].

An attractive strategy embraced by worldwide practitioners in order to provide additional bactericidal effects to general-use dental materials is to modify or embed them with silver-based nanostructures [150]. Though silver has favorable effects in caries prophylaxis in the form of nanosilver diamine fluoride (SDF), the use of this particular compound has some disadvantages, one of the most noticeable effects being represented by tooth staining [151]. By reducing the size of AgNPs, the contact surface will be considerably increased; in this way, the antimicrobial effects of silver would be improved, and the use of nanosilver could prevent black staining in teeth, which usually occurs after the application of SDF [152].

Antibacterial resins could be used in clinical dental applications, both in orthodontics and restorative dentistry [153]. In orthodontics, these resins could be used as bracket or branked bonding materials, while, in restorative dentistry, they could be used as filling or denture base material [153]. Therefore, in order to improve their physico-mechanical properties and antimicrobial effects, a method for incorporating AgNPs into acrylic resin denture-base materials was developed [154].

Because the oral cavity is an active ecosystem usually colonized by various pathogenic microorganisms, dental materials and implants have an increased risk of contamination and subsequent colonization processes [155]. In terms of superior antimicrobial activity, promising results were reported with respect to the incorporation of silver-based nanosystems within adhesive resins [156,157], orthodontic cements [158,159], and dental composites [160–162]. In addition to being used as antimicrobial filling agents within multifunctional biomaterials, another attractive and challenging dental application of AgNPs relies on their potential use as biostatic or biocide coatings for conventional titanium-based dental implants [163,164]. Though AgNPs proved to be efficient and effective agents in dental practice, they remain controversial candidates for this specific area of research, due to their variable toxicity in biological systems. Therefore, any potential application of AgNPs in dentistry must include thorough studies regarding the optimal compromise between physicochemical features and biofunctional performance [165].

6. Silver Nanoparticles for Wound Healing

Wound infections represent an important clinical challenge, with major impact on patient morbidity and mortality and notable economic implications [166]. Preventing wound dehiscence and surgical-site infection are challenging and essential aspects in current clinical practice [167]. The skin is the most extensive and one of the most complex organs in the human body, but it can be easily affected by different harmful external factors [168]. Physically or chemically induced cutaneous wounds may significantly disturb skin structural and functional integrity at different stages, leading to permanent disability or even death, depending on the severity of the injury [169]. In the past few years, wound infections caused by opportunistic pathogenic microorganism became an important issue during current healthcare practice [170]. The ultimate tendency and ideal desideratum for infected-wound management is represented by fast tissue-recovery processes, accompanied by maximal functionality restoration and minimal scar-tissue formation [171]. The wound-healing process, as any complex pathophysiological mechanism, includes different stages, such as coagulation, inflammation, cellular proliferation, and matrix and tissue remodeling [171].

Since ancient times, silver-based compounds and materials were used for the unconventional and effective control of distinctive infections [172]. Given its intrinsic physicochemical features and biological peculiarities, nanosilver provides a wide range of efficient biocide activities against an impressive diversity of anaerobic and aerobic, Gram-negative and Gram-positive bacterial strains. It is well known that bacterial and mammalian cells poorly absorb metallic or elemental silver, due to its chemical inactivation. Therefore, in order to provide specific antibacterial effects under physiological conditions (including the presence of body fluids or secretions), the ionization of silver is required. After their penetration inside cells, silver ions merge with enzymatic and structural proteins [173]. AgNPs or silver ions used in absorbent wound dressings can interact with and destroy the bacteria found in exudate [174].

Briefly, recent data provide the following information regarding AgNP skin absorption: (i) there is plenty experimental evidence with respect to the in vitro skin permeation by nanoparticles, and (ii) there is an important increase in permeation in the case of damaged skin [175]. When naturally available biopolymers (e.g., chitosan [176] or collagen [177]) are implied in novel nanotechnology approaches, they possess tremendous potential regarding the obtaining of novel and functionally improved platforms for effective wound-healing applications [178].

Acticoat™ and Bactigras™ (Smith & Nephew), Aquacel™ (ConvaTec), PolyMem Silver™ (Aspen), and Tegaderm™ (3M) are representative biocomposites modified with ionic silver and approved by the United States (US) Food and Drug Administration (FDA) for wound-dressing applications. In addition to these commercial products, promising results were reported with respect to the incorporation of AgNPs within novel and naturally derived biomaterials for enhanced wound-healing management, such as (but not limited to) modified cotton fabrics [179,180], bacterial cellulose [181,182], chitosan [176,183], and sodium alginate [184,185].

The use of AgNPs and Ag^+ carriers also represents a valuable strategy for delayed diabetic wound-healing processes, since diabetic wounds may be accompanied by numerous secondary infections. AgNPs can help diabetic patients in early wound-healing stages, additionally providing minor scars [186]. Taking into account the efficient and enhanced antibacterial effects exhibited by AgNPs and the impressive interest oriented toward their application in wound therapy and medical-device coatings, their biocompatibility and safety aspects must be thoroughly clarified [187].

7. Silver Nanoparticles for Bone Healing

Every year, millions of people worldwide are affected by distinctive and complex bone-related pathologies, including infectious diseases, degenerative and genetic conditions, cancers, and fractures [188]. Unfortunately, the opportunistic contamination and colonization of orthopedic implants represent major concerns in osseous-tissue replacement strategies, since the related infections are associated with high morbidity [189]. Bone is an active tissue that undergoes regenerative and restorative processes through the intrinsic and complex bone-remodeling mechanism [190]. Bone grafts are usually implanted to replace or restore severe defects that irremediably affect osseous tissue, such as genetic malformations, tumors, or traumas [191]. Orthopedic and bone-implant-related infections are usually associated with highly inflammatory processes and subsequent implant loss and bone-destruction phenomena [192].

Previous studies reported that AgNPs naturally improve the differentiation process of MC3T3-1 pre-osteoblast cells and subsequent bone-like tissue mineralization, when compared with other NPs [193]. Currently, silver-coated prostheses represent an unconventional approach during the prophylaxis of tumor-related infections and extensive trauma-related infections. However, no clinical studies comparing the long-term clinical impact of nanosilver-coated implants for revision arthroplasty are reported as of yet [194]. The self-repairing capability of bone can be limited when bacterial activity occurs in bone defects. Compared with usual antibiotics, AgNPs possess intrinsic antibacterial activity with a broader spectrum. Also, the bacterial resistance to AgNP activity is an uncommon phenomenon, thus emphasizing that the bactericidal mechanisms of nanosilver act in synergy. Thanks to this peculiar property, AgNPs have the capability to inhibit or impair biofilm development or mature biofilm, respectively, in the case of antibiotic-resistant bacteria, such as methicillin-resistant *Staphylococcus aureus* [195].

Human bone, dentin, and dental enamel are mainly composed of crystallized hydroxyapatite (HA), which is a calcium-phosphate salt [196]. Given the specific biocompatibility of biosynthesized and synthetic HA, this material and its derivatives are extensively explored for the development of unconventional osseous-related restorative and regenerative strategies, either as artificial bone grafts or as coating materials for metallic implants [197]. With regards to the superficial modification of various metallic implant surfaces, biocompatible HA integrated with silver (either in metallic or ionic form) represents a suitable choice for the fabrication of bioactive and antimicrobial bone implants [198]. The antimicrobial efficiency of HA-based coatings embedded with nanosilver was evidenced against Gram-positive [199–201] and Gram-negative [202–204] bacterial strains.

In terms of bone-replacement procedures, AgNPs are normally used as doping materials for synthetic and bio-inspired bone scaffolds, with relevant results being recently reported [205,206]. In order to induce antibacterial properties in HA coatings, several experimental techniques proved suitable for the incorporation of nanosilver within calcium-phosphate materials, such as laser-assisted deposition, electrochemical deposition, magnetron sputtering, ion-beam-assisted deposition, sol-gel technology, and microarc oxidation [207].

Previous studies showed that AgNP-implanted titanium displayed improved antibacterial ability, as well as excellent compatibility with osteoblasts, thanks to the micro-galvanic effects produced between the implanted AgNPs and the titanium substrate [208]. Many studies investigated the feasibility and clinical potential of adjusting acrylic cements with AgNPs, in order to provide unconventional and functionally improved biomaterials for orthopedic applications. While previous studies explored different acrylics modified with AgNPs, a significant part of the previous work is limited, since vital material characteristics and mechanical properties were not thoroughly analyzed [159,209–211].

Moreover, the beneficial addition of antimicrobial AgNPs within composite matrices designed for bone-tissue engineering were emphasized. In a recent study, it was shown that AgNPs could promote the osteogenesis and proliferation of mesenchymal stem cells (MSCs), in order to enhance the healing process of bone fracture [212]. A correlation was also reported between NP uptake and growth in clathrin-dependent endocytosis in the case of MSCs and osteoblasts, indicating that this route may represent the principal cellular internalization pathway of AgNPs [213]. Taking into account the limited capacity of bone tissue to fully reconstruct or replace severe defects, the development of novel and performance-enhanced implants is required. Thus, new pathways were used to stimulate bone regeneration and also to prevent the side effects correlated with therapeutics currently used in the clinic [214].

8. Silver Nanoparticles for Other Medical Applications

Thanks to their unique physiochemical properties and biofunctional features, such as anti-inflammatory, anti-angiogenesis, antiplatelet, antiviral, antifungal, and antibacterial activities, AgNPs play an important role in the development and implementation of novel biomedicinal strategies [45]. Recently, AgNPs were intimately investigated regarding their promising anticancer effects exhibited in different human cancerous cell lines, such as endothelial cells, IMR-90 lung fibroblasts, U251 glioblastoma cells, and MDA-MB-231 breast cancer cells [215,216]. AgNPs possess the intrinsic capability to merge with mammalian cells and to easily penetrate them by means of energy-driven internalization pathways [217]. Another attractive property of AgNPs relies on their specific fluorescence, making them suitable candidates for detection and dose-enhancement purposes in X-ray irradiation applications [218].

At the moment, the combination of therapy and diagnosis, known as theranostics, represents the most important, attractive, and challenging approach embraced by healthcare practitioners and researchers with respect to the effective and personalized therapy of cancer desideratum [219]. AgNPs are also plasmonic structures, capable of particularly scattering and absorbing the light impinging certain areas. After their selective uptake into cancerous cells, AgNP-derived scattered light can be used for imaging purposes, whereas absorbed light can be used for selective hyperthermia [220].

Cardiovascular diseases (CVDs) represent a major cause of worldwide human death, being responsible for more than 17.7 million deaths in 2015 [221]. Recently, many studies focused on the evaluation of the effects of AgNPs on various types of cell encountered in the complex vascular system, but the reported results were contradictory. However, the collected data can provide substantial knowledge with respect to the potential benefits of AgNPs for pathological and physiological stages related to the cardiovascular system, thus contributing to the development of novel and specific molecular therapies in vascular tone, vasopermeability, and angiogenesis [222]. Cardiovascular pathologies, such as hypertension, may influence the toxic effects induced by AgNPs [223]. The first silver-modified cardiovascular medical device was a prosthetic silicone heart valve coated with elemental silver, which was developed to avoid valve-related bacterial infection and to reduce inflammation response [224].

Malaria, one of the most common infectious diseases encountered in tropical and sub-tropical regions, became a major healthcare concern all around the world. It was shown that AgNPs possess powerful activity against both the malarial parasite (*Plasmodium falciparum*) and its related vector (*Anopheles* female mosquito). The intrinsic anti-plasmodial effects exhibited by nanosilver-based compounds and materials represent a solid starting point toward the nanotechnology-derived therapy and worldwide control of malaria [24,225,226].

The human eye is a complex organ, with impressive vascularization and innervation, that can be easily exposed to microbial contamination under proper temperature and humidity conditions [227,228]. Nanosilver-based compounds and materials proved promising potential toward the development of unconventional and performance-enhanced therapy of eye-related infectious conditions. AgNPs coated with calcium indicators proved to have reduced damage with respect to retinal cells, and could be experimentally applied for retinal imaging in a mouse animal model [229,230]. The bactericidal effects related to AgNP-containing nanomaterials are essential aspects which must be further considered for their exploitation as an improved class of antibacterial agent for ocular applications [14,231,232].

AgNPs can be successfully used as novel nanostructured platforms for diagnostics and the treatment of different cancers [233]. The broad-spectrum bioactivity of AgNPs makes them promising agents not only for anti-infective fighting strategies, but also in critical tumor and multi-drug resistance tackling approaches.

9. Toxicity of Silver Nanoparticles

Even if AgNPs possess tremendous advantages that recommend them for novel and challenging biomedical applications, their toxicity became an intensive study subject only recently. The daily amount of silver derived from natural sources in food and water ingested by humans is approximately 0.4–30 μg [234]. The available studies performed with respect to the toxic effects exhibited by AgNPs within biological systems, such as bacteria and viruses or human cells, report contradictory and various results [235,236]. AgNPs are generally presented as highly effective antimicrobial agents with non-toxic effects to healthy mammalian cells [237]. However, various in vitro studies demonstrated the nanosilver-related toxic effects in rat hepatocytes and neuronal cells [238], murine stem cells, and human lung epithelial cells [239]. The toxicity of AgNPs was also investigated during in vivo assays. The toxicity studies performed in a rat ear model proved that AgNP exposure resulted in significant mitochondrial dysfunction and subsequent temporary or permanent hearing loss, depending on the inoculation dose. Even low concentrations of AgNPs were absorbed by retinal cells and resulted in important structural disruption, due to the increased number of cells that underwent oxidative stress [240].

The possible toxicity mechanisms related to AgNPs are depicted in Figure 2 [241]. The performed studies also proved that variations in surface charge resulting from the surface functionalization of AgNPs can impact cellular uptake, translocation to various tissues, and cytotoxicity. The magnitude of the surface charge, as measured by the zeta potential, can influence the amount of nanoparticles and their mechanism of uptake into cells [242].

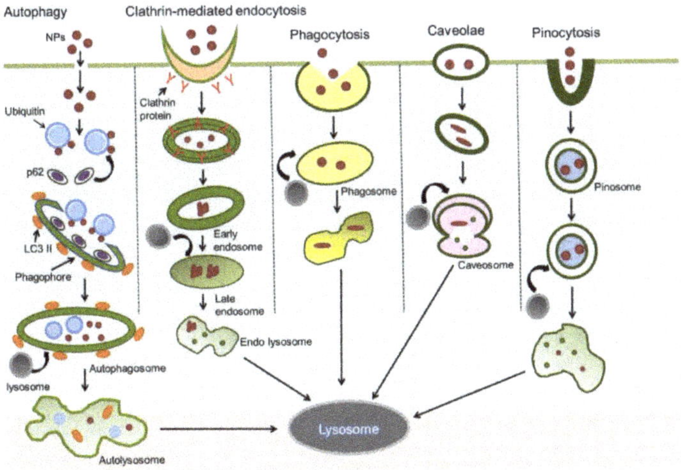

Figure 2. Schematic representation of plausible methods of cellular uptake of AgNPs [243]. Reprinted with permission from [243]. Elsevier, 2015.

In order to investigate the toxic effects caused by exposure to nanosilver-based systems, thorough assays are required, considering both cellular and animal models. Regarding the in vivo biocompatibility and biodistribution assays, the reported data evidenced that AgNPs can result in structural and physiological alteration of vital organs. For example, inhaled AgNPs may form deposits in the alveolar regions, leading to lung injuries, and may also generate significant modifications within the nervous system, and liver and kidney tissues. Intratracheal instillation of AgNPs can affect vascular reactivity and can further exacerbate cardiac reperfusion/ischemia injury [244,245].

The toxicity of AgNPs is related to their transformation under biological conditions and environmental media, including their interactions with biological macromolecules, surface oxidation, and the release of silver ions. Also, it is very important to precisely distinguish the toxicity rate related to either nanosized silver or ionic silver [246]. Many studies proved that AgNP exposure can induce a decrease in cell viability through different cellular mechanisms. One of these mechanisms is represented by the induction of apoptosis-related genes and the activation of apoptosis mechanism. Also, it was proven that nanosilver can cause the formation and intracellular accumulation of ROS, modification of mitochondrial membrane permeability, and DNA damage [247–249]. The in vitro toxicity of AgNPs was investigated in several research studies, but there is still a lack of consistent and reliable data. This is a general concern in nanotoxicology, and more research coherence is needed to produce meaningful results. According to recent data, the main in vitro outcomes occurring upon exposure to AgNPs were reported as increases in oxidative stress, genotoxicity, and apoptosis levels [250–252]. AgNPs may induce significant oxidative damage with respect to the cellular membrane and organelles such as the nucleus, mitochondria, and lysosomes, thus leading directly to necrotic or apoptotic phenomena. The oxidative stress caused by AgNPs can result in inflammatory responses, including the activation of innate immunity and increasing the permeability of endothelial cells [253]. AgNPs, inoculated at non-cytotoxic doses, may cause chromosomal abnormality, DNA damage, and possible mutagenicity [254–256].

The genotoxicity and cytotoxicity of AgNPs are influenced by several physicochemical features, including dispersion rate, concentration, surface charge, size, morphology, and surface functionalization [257,258]. The physicochemical aspects of nanosilver-based systems and materials mainly distribute and categorize numerous toxicological concerns, and also establish a ladder of toxicity framework while imposing on the biological system. The experimental results reported until recently are insufficient regarding the accurate toxic effects of AgNPs and their related toxicity mechanisms [35,259].

10. Conclusions

Silver nanoparticles (AgNPs) are intensively explored nanostructures for unconventional and enhanced biomedical applications, thanks to their size-related attractive physicochemical properties and biological functionality, including their high antimicrobial efficiency and non-toxic nature. AgNP-based nanosystems and nanomaterials are suitable alternatives for drug delivery, wound dressing, tissue scaffold, and protective coating applications. Various physicochemical parameters were related to the intrinsic antimicrobial effects exhibited by AgNPs, such as size, shape, concentration, surface charge, and colloidal state. Moreover, the impressive available surface of nanosilver allows the coordination of many ligands, thus enabling tremendous possibilities with respect to the surface functionalization of AgNPs.

There is a significant amount of research data proving the beneficial effects of AgNPs in novel biocompatible and nanostructured materials and devices developed for modern therapeutic strategies. In addition to their attractive and versatile antimicrobial potential, AgNPs provide additional mechanical, optical, chemical, and biological peculiarities that recommend them for the design, obtaining, evaluation, and clinical assessment of performance-enhanced biomaterials and medical devices. Still, thorough investigations regarding their short-term and long-term toxicity, as well as the responsible toxic-related mechanisms, are required.

The current limitations related to conventional healthcare practice and the latest challenges resulting from nanosilver-based technology outline the impressive potential of silver nanoparticles in biomedicinal applications. Whether we consider the modification of available biomaterials and devices or the development of novel nanostructured ones, AgNPs are ideal candidates for achieving the very close modern biomedicine desideratum.

Author Contributions: A.-C.B., O.G., A.M.G., L.M., A.F., and E.A. designed and wrote the paper.

Acknowledgments: This work was supported by a grant from the Romanian National Authority for Scientific Research and Innovation, UEFISCDI, project number 45PCCDI/2018-Nanostructuri bioactive pentru strategii terapeutice inovatoare.

Conflicts of Interest: The authors declare no conflict of interest.

References

1. Abou El-Nour, K.M.M.; Eftaiha, A.; Al-Warthan, A.; Ammar, R.A.A. Synthesis and applications of silver nanoparticles. *Arab. J. Chem.* **2010**, *3*, 135–140. [CrossRef]
2. Faisal, N.; Kumar, K. Polymer and metal nanocomposites in biomedical applications. *Biointerface Res. Appl. Chem.* **2017**, *7*, 2286–2294.
3. Alexander, J.W. History of the medical use of silver. *Surg. Infect.* **2009**, *10*, 289–292. [CrossRef] [PubMed]
4. Ioan-Avram, N.; Anton, F.; Maria, S.; Denisa, F.; Ovidiu, O.; Ecaterina, A. Silver based materials for biomedical applications. *Curr. Org. Chem.* **2014**, *18*, 173–184.
5. Geraldo, D.A.; Needham, P.; Chandia, N.; Arratia-Perez, R.; Mora, G.C.; Villagra, N.A. Green synthesis of polysaccharides-based gold and silver nanoparticles and their promissory biological activity. *Biointerface Res. Appl. Chem.* **2016**, *6*, 1263–1271.
6. Chowdhury, N.R.; MacGregor-Ramiasa, M.; Zilm, P.; Majewski, P.; Vasilev, K. 'Chocolate' silver nanoparticles: Synthesis, antibacterial activity and cytotoxicity. *J. Colloid Interface Sci.* **2016**, *482*, 151–158. [CrossRef] [PubMed]
7. Tavaf, Z.; Tabatabaei, M.; Khalafi-Nezhad, A.; Panahi, F. Evaluation of antibacterial, antibofilm and antioxidant activities of synthesized silver nanoparticles (AgNPs) and casein peptide fragments against streptococcus mutans. *Eur. J. Integr. Med.* **2017**, *12*, 163–171. [CrossRef]
8. Domeradzka-Gajda, K.; Nocun, M.; Roszak, J.; Janasik, B.; Quarles, C.D., Jr.; Wasowicz, W.; Grobelny, J.; Tomaszewska, E.; Celichowski, G.; Ranoszek-Soliwoda, K.; et al. A study on the in vitro percutaneous absorption of silver nanoparticles in combination with aluminum chloride, methyl paraben or di-n-butyl phthalate. *Toxicol. Lett.* **2017**, *272*, 38–48. [CrossRef] [PubMed]
9. Kraeling, M.E.K.; Topping, V.D.; Keltner, Z.M.; Belgrave, K.R.; Bailey, K.D.; Gao, X.; Yourick, J.J. In vitro percutaneous penetration of silver nanoparticles in pig and human skin. *Regul. Toxicol. Pharm.* **2018**, *95*, 314–322. [CrossRef] [PubMed]
10. Fortunati, E.; Peltzer, M.; Armentano, I.; Jiménez, A.; Kenny, J.M. Combined effects of cellulose nanocrystals and silver nanoparticles on the barrier and migration properties of pla nano-biocomposites. *J. Food Eng.* **2013**, *118*, 117–124. [CrossRef]
11. Kumar, S.; Shukla, A.; Baul, P.P.; Mitra, A.; Halder, D. Biodegradable hybrid nanocomposites of chitosan/gelatin and silver nanoparticles for active food packaging applications. *Food Packag. Shelf* **2018**, *16*, 178–184. [CrossRef]
12. Pannerselvam, B.; Dharmalingam Jothinathan, M.K.; Rajenderan, M.; Perumal, P.; Pudupalayam Thangavelu, K.; Kim, H.J.; Singh, V.; Rangarajulu, S.K. An in vitro study on the burn wound healing activity of cotton fabrics incorporated with phytosynthesized silver nanoparticles in male Wistar albino rats. *Eur. J. Pharm. Sci.* **2017**, *100*, 187–196. [CrossRef] [PubMed]
13. Zhou, Y.; Tang, R.C. Facile and eco-friendly fabrication of agnps coated silk for antibacterial and antioxidant textiles using honeysuckle extract. *J. Photochem. Photobiol. B* **2018**, *178*, 463–471. [CrossRef] [PubMed]
14. Kejlová, K.; Kašpárková, V.; Krsek, D.; Jírová, D.; Kolářová, H.; Dvořáková, M.; Tománková, K.; Mikulcová, V. Characteristics of silver nanoparticles in vehicles for biological applications. *Int. J. Pharm.* **2015**, *496*, 878–885. [CrossRef] [PubMed]
15. Zhang, L.; Zeng, G.; Dong, H.; Chen, Y.; Zhang, J.; Yan, M.; Zhu, Y.; Yuan, Y.; Xie, Y.; Huang, Z. The impact of silver nanoparticles on the co-composting of sewage sludge and agricultural waste: Evolutions of organic matter and nitrogen. *Bioresour. Technol.* **2017**, *230*, 132–139. [CrossRef] [PubMed]

16. Gupta, S.D.; Agarwal, A.; Pradhan, S. Phytostimulatory effect of silver nanoparticles (AgNPs) on rice seedling growth: An insight from antioxidative enzyme activities and gene expression patterns. *Ecotoxicol. Environ. Saf.* **2018**, *161*, 624–633. [CrossRef] [PubMed]
17. Achmad, S.; Salmiati; Razman, S.M.; Ahmad, B.H.K.; Tony, H.; Hadi, N. A review of silver nanoparticles: Research trends, global consumption, synthesis, properties, and future challenges. *J. Chin. Chem. Soc.* **2017**, *64*, 732–756.
18. Brobbey, K.J.; Haapanen, J.; Gunell, M.; Mäkelä, J.M.; Eerola, E.; Toivakka, M.; Saarinen, J.J. One-step flame synthesis of silver nanoparticles for roll-to-roll production of antibacterial paper. *Appl. Surf. Sci.* **2017**, *420*, 558–565. [CrossRef]
19. He, R.; Ren, F.; Chen, F. Embedded silver nanoparticles in ktp crystal produced by ion implantation. *Mater. Lett.* **2017**, *193*, 158–160. [CrossRef]
20. Han, H.J.; Yu, T.; Kim, W.-S.; Im, S.H. Highly reproducible polyol synthesis for silver nanocubes. *J. Cryst. Growth* **2017**, *469*, 48–53. [CrossRef]
21. Khatoon, U.T.; Nageswara Rao, G.V.S.; Mohan, K.M.; Ramanaviciene, A.; Ramanavicius, A. Antibacterial and antifungal activity of silver nanospheres synthesized by tri-sodium citrate assisted chemical approach. *Vacuum* **2017**, *146*, 259–265. [CrossRef]
22. Verma, S.; Rao, B.T.; Srivastava, A.P.; Srivastava, D.; Kaul, R.; Singh, B. A facile synthesis of broad plasmon wavelength tunable silver nanoparticles in citrate aqueous solutions by laser ablation and light irradiation. *Colloids Surf. A Physicochem. Eng. Asp.* **2017**, *527*, 23–33. [CrossRef]
23. Liu, F.; Liu, J.; Cao, X. Microwave-assisted synthesis silver nanoparticles and their surface enhancement raman scattering. *Rare Met. Mater. Eng.* **2017**, *46*, 2395–2398.
24. Dutta, P.P.; Bordoloi, M.; Gogoi, K.; Roy, S.; Narzary, B.; Bhattacharyya, D.R.; Mohapatra, P.K.; Mazumder, B. Antimalarial silver and gold nanoparticles: Green synthesis, characterization and in vitro study. *Biomed. Pharmacother.* **2017**, *91*, 567–580. [CrossRef] [PubMed]
25. Singh, T.; Jyoti, K.; Patnaik, A.; Singh, A.; Chauhan, R.; Chandel, S.S. Biosynthesis, characterization and antibacterial activity of silver nanoparticles using an endophytic fungal supernatant of raphanus sativus. *J. Genet. Eng. Biotechnol.* **2017**, *15*, 31–39. [CrossRef]
26. Mahmoud, K.H.; Abbo, M. Synthesis, characterization and optical properties of gelatin doped with silver nanoparticles. *Spectrochim. Acta Part A Mol. Biomol. Spectrosc.* **2013**, *116*, 610–615. [CrossRef] [PubMed]
27. Hanif, M.; Juluri, R.R.; Fojan, P.; Popok, V.N. Polymer films with size-selected silver nanoparticles as plasmon resonance-based transducers for protein sensing. *Biointerface Res. Appl. Chem.* **2016**, *6*, 1564–1568.
28. Higa, A.M.; Mambrini, G.P.; Hausen, M.; Strixino, F.T.; Leite, F.L. Ag-nanoparticle-based nano-immunosensor for anti-glutathione s-transferase detection. *Biointerface Res. Appl. Chem.* **2016**, *6*, 1053–1058.
29. Qiu, C.; Bennet, K.E.; Tomshine, J.R.; Hara, S.; Ciubuc, J.D.; Schmidt, U.; Durrer, W.G.; McIntosh, M.B.; Eastman, M.; Manciu, F.S. Ultrasensitive detection of neurotransmitters by surface enhanced raman spectroscopy for biosensing applications. *Biointerface Res. Appl. Chem.* **2017**, *7*, 1921–1926.
30. Chien, C.-S.; Lin, C.-J.; Ko, C.-J.; Tseng, S.-P.; Shih, C.-J. Antibacterial activity of silver nanoparticles (AgNP) confined to mesostructured silica against methicillin-resistant staphylococcus aureus (MRSA). *J. Alloys Compd.* **2018**, *747*, 1–7. [CrossRef]
31. Lau, C.P.; Abdul-Wahab, M.F.; Jaafar, J.; Chan, G.F.; Abdul Rashid, N.A. Toxic effect of high concentration of sonochemically synthesized polyvinylpyrrolidone-coated silver nanoparticles on citrobacter sp. A1 and enterococcus sp. C1. *J. Microbiol. Immunol. Infect.* **2017**, *50*, 427–434. [CrossRef] [PubMed]
32. Muhammad, Z.; Raza, A.; Ghafoor, S.; Naeem, A.; Naz, S.S.; Riaz, S.; Ahmed, W.; Rana, N.F. Peg capped methotrexate silver nanoparticles for efficient anticancer activity and biocompatibility. *Eur. J. Pharm. Sci.* **2016**, *91*, 251–255. [CrossRef] [PubMed]
33. Panzarini, E.; Mariano, S.; Vergallo, C.; Carata, E.; Fimia, G.M.; Mura, F.; Rossi, M.; Vergaro, V.; Ciccarella, G.; Corazzari, M.; et al. Glucose capped silver nanoparticles induce cell cycle arrest in hela cells. *Toxicol. In Vitro* **2017**, *41*, 64–74. [CrossRef] [PubMed]
34. He, H.; Tao, G.; Wang, Y.; Cai, R.; Guo, P.; Chen, L.; Zuo, H.; Zhao, P.; Xia, Q. In situ green synthesis and characterization of sericin-silver nanoparticle composite with effective antibacterial activity and good biocompatibility. *Mater. Sci. Eng. C* **2017**, *80*, 509–516. [CrossRef] [PubMed]

35. Akter, M.; Sikder, M.T.; Rahman, M.M.; Ullah, A.K.M.A.; Hossain, K.F.B.; Banik, S.; Hosokawa, T.; Saito, T.; Kurasaki, M. A systematic review on silver nanoparticles-induced cytotoxicity: Physicochemical properties and perspectives. *J. Adv. Res.* **2018**, *9*, 1–16. [CrossRef] [PubMed]
36. Zheng, K.; Setyawati, M.I.; Leong, D.T.; Xie, J. Antimicrobial silver nanomaterials. *Coord. Chem. Rev.* **2018**, *357*, 1–17. [CrossRef]
37. Durán, N.; Durán, M.; de Jesus, M.B.; Seabra, A.B.; Fávaro, W.J.; Nakazato, G. Silver nanoparticles: A new view on mechanistic aspects on antimicrobial activity. *Nanomed. Nanotechnol. Biol. Med.* **2016**, *12*, 789–799. [CrossRef] [PubMed]
38. Thomas, R.; Soumya, K.R.; Mathew, J.; Radhakrishnan, E.K. Inhibitory effect of silver nanoparticle fabricated urinary catheter on colonization efficiency of coagulase negative staphylococci. *J. Photochem. Photobiol. B Biol.* **2015**, *149*, 68–77. [CrossRef] [PubMed]
39. Marassi, V.; Di Cristo, L.; Smith, S.G.J.; Ortelli, S.; Blosi, M.; Costa, A.L.; Reschiglian, P.; Volkov, Y.; Prina-Mello, A. Silver nanoparticles as a medical device in healthcare settings: A five-step approach for candidate screening of coating agents. *R. Soc. Open Sci.* **2018**, *5*, 171113. [CrossRef] [PubMed]
40. Koduru, J.R.; Kailasa, S.K.; Bhamore, J.R.; Kim, K.-H.; Dutta, T.; Vellingiri, K. Phytochemical-assisted synthetic approaches for silver nanoparticles antimicrobial applications: A review. *Adv. Colloid Interface Sci.* **2018**, *256*, 326–339. [CrossRef] [PubMed]
41. Alshareef, A.; Laird, K.; Cross, R.B.M. Shape-dependent antibacterial activity of silver nanoparticles on *Escherichia coli* and *Enterococcus faecium* bacterium. *Appl. Surf. Sci.* **2017**, *424*, 310–315. [CrossRef]
42. Adur, A.J.; Nandini, N.; Shilpashree Mayachar, K.; Ramya, R.; Srinatha, N. Bio-synthesis and antimicrobial activity of silver nanoparticles using anaerobically digested parthenium slurry. *J. Photochem. Photobiol. B Biol.* **2018**, *183*, 30–34. [CrossRef] [PubMed]
43. Etemadzade, M.; Ghamarypour, A.; Zabihollahi, R.; shabbak, G.; Shirazi, M.; Sahebjamee, H.; Vaziri, A.Z.; Assadi, A.; Ardestani, M.S.; Shandiz, S.A.S.; et al. Synthesis and evaluation of antiviral activities of novel sonochemical silver nanorods against hiv and hsv viruses. *Asian Pac. J. Trop. Dis.* **2016**, *6*, 854–858. [CrossRef]
44. Tamilselvan, S.; Ashokkumar, T.; Govindaraju, K. Microscopy based studies on the interaction of bio-based silver nanoparticles with bombyx mori nuclear polyhedrosis virus. *J. Virol. Methods* **2017**, *242*, 58–66. [CrossRef] [PubMed]
45. Kalaivani, R.; Maruthupandy, M.; Muneeswaran, T.; Hameedha Beevi, A.; Anand, M.; Ramakritinan, C.M.; Kumaraguru, A.K. Synthesis of chitosan mediated silver nanoparticles (Ag NPs) for potential antimicrobial applications. *Front. Lab. Med.* **2018**, *2*, 30–35. [CrossRef]
46. Dojčilović, R.; Pajović, J.D.; Božanić, D.K.; Bogdanović, U.; Vodnik, V.V.; Dimitrijević-Branković, S.; Miljković, M.G.; Kaščaková, S.; Réfrégiers, M.; Djoković, V. Interaction of amino acid-functionalized silver nanoparticles and Candida albicans polymorphs: A deep-UV fluorescence imaging study. *Colloids Surf. B Biointerfaces* **2017**, *155*, 341–348. [CrossRef] [PubMed]
47. Zhang, X.F.; Liu, Z.G.; Shen, W.; Gurunathan, S. Silver nanoparticles: Synthesis, characterization, properties, applications, and therapeutic approaches. *Int. J. Mol. Sci.* **2016**, *17*, 1534. [CrossRef] [PubMed]
48. Wei, L.; Lu, J.; Xu, H.; Patel, A.; Chen, Z.S.; Chen, G. Silver nanoparticles: Synthesis, properties, and therapeutic applications. *Drug Discov. Today* **2015**, *20*, 595–601. [CrossRef] [PubMed]
49. Singh, P.; Kim, Y.J.; Singh, H.; Wang, C.; Hwang, K.H.; Farh, M.E.-A.; Yang, D.C. Biosynthesis, characterization, and antimicrobial applications of silver nanoparticles. *Int. J. Nanomed.* **2015**, *10*, 2567–2577.
50. Mokhena, T.C.; Luyt, A.S. Electrospun alginate nanofibres impregnated with silver nanoparticles: Preparation, morphology and antibacterial properties. *Carbohydr. Polym.* **2017**, *165*, 304–312. [CrossRef] [PubMed]
51. Gudikandula, K.; Vadapally, P.; Singara Charya, M.A. Biogenic synthesis of silver nanoparticles from white rot fungi: Their characterization and antibacterial studies. *OpenNano* **2017**, *2*, 64–78. [CrossRef]
52. Guan, Q.; Xia, C.; Li, W. Bio-friendly controllable synthesis of silver nanoparticles and their enhanced antibacterial property. *Catal. Today* **2018**. [CrossRef]
53. Li, W.-R.; Sun, T.-L.; Zhou, S.-L.; Ma, Y.-K.; Shi, Q.-S.; Xie, X.-B.; Huang, X.-M. A comparative analysis of antibacterial activity, dynamics, and effects of silver ions and silver nanoparticles against four bacterial strains. *Int. Biodeterior. Biodegrad.* **2017**, *123*, 304–310. [CrossRef]

54. Premkumar, J.; Sudhakar, T.; Dhakal, A.; Shrestha, J.B.; Krishnakumar, S.; Balashanmugam, P. Synthesis of silver nanoparticles (AgNPs) from cinnamon against bacterial pathogens. *Biocatal. Agric. Biotechnol.* **2018**, *15*, 311–316. [CrossRef]
55. Shao, Y.; Wu, C.; Wu, T.; Yuan, C.; Chen, S.; Ding, T.; Ye, X.; Hu, Y. Green synthesis of sodium alginate-silver nanoparticles and their antibacterial activity. *Int. J. Biol. Macromol.* **2018**, *111*, 1281–1292. [CrossRef] [PubMed]
56. Yan, X.; He, B.; Liu, L.; Qu, G.; Shi, J.; Hu, L.; Jiang, G. Antibacterial mechanism of silver nanoparticles in pseudomonas aeruginosa: Proteomics approach. *Metallomics* **2018**, *10*, 557–564. [CrossRef] [PubMed]
57. Prabhu, S.; Poulose, E.K. Silver nanoparticles: Mechanism of antimicrobial action, synthesis, medical applications, and toxicity effects. *Int. Nano Lett.* **2012**, *2*, 32. [CrossRef]
58. López-Esparza, J.; Espinosa-Cristóbal, L.F.; Donohue-Cornejo, A.; Reyes-López, S.Y. Antimicrobial activity of silver nanoparticles in polycaprolactone nanofibers against gram-positive and gram-negative bacteria. *Ind. Eng. Chem. Res.* **2016**, *55*, 12532–12538. [CrossRef]
59. Bhat, R.; Deshpande, R.; Ganachari, S.V.; Huh, D.S.; Venkataraman, A. Photo-irradiated biosynthesis of silver nanoparticles using edible mushroom pleurotus florida and their antibacterial activity studies. *Bioinorg. Chem. Appl.* **2011**, *2011*, 650979. [CrossRef] [PubMed]
60. Izak-Nau, E.; Huk, A.; Reidy, B.; Uggerud, H.; Vadset, M.; Eiden, S.; Voetz, M.; Himly, M.; Duschl, A.; Dusinska, M.; et al. Impact of storage conditions and storage time on silver nanoparticles' physicochemical properties and implications for their biological effects. *RSC Adv.* **2015**, *5*, 84172–84185. [CrossRef]
61. Lee, J.-H.; Lim, J.-M.; Velmurugan, P.; Park, Y.-J.; Park, Y.-J.; Bang, K.-S.; Oh, B.-T. Photobiologic-mediated fabrication of silver nanoparticles with antibacterial activity. *J. Photochem. Photobiol. B Biol.* **2016**, *162*, 93–99. [CrossRef] [PubMed]
62. Ghiuţă, I.; Cristea, D.; Croitoru, C.; Kost, J.; Wenkert, R.; Vyrides, I.; Anayiotos, A.; Munteanu, D. Characterization and antimicrobial activity of silver nanoparticles, biosynthesized using bacillus species. *Appl. Surf. Sci.* **2018**, *438*, 66–73. [CrossRef]
63. De Faria, A.F.; Martinez, D.S.T.; Meira, S.M.M.; de Moraes, A.C.M.; Brandelli, A.; Filho, A.G.S.; Alves, O.L. Anti-adhesion and antibacterial activity of silver nanoparticles supported on graphene oxide sheets. *Colloids Surf. B Biointerfaces* **2014**, *113*, 115–124. [CrossRef] [PubMed]
64. Zhou, Y.; Hu, K.; Guo, Z.; Fang, K.; Wang, X.; Yang, F.; Gu, N. Plla microcapsules combined with silver nanoparticles and chlorhexidine acetate showing improved antibacterial effect. *Mater. Sci. Eng. C Mater. Biol. Appl.* **2017**, *78*, 349–353. [CrossRef] [PubMed]
65. Amooaghaie, R.; Saeri, M.R.; Azizi, M. Synthesis, characterization and biocompatibility of silver nanoparticles synthesized from nigella sativa leaf extract in comparison with chemical silver nanoparticles. *Ecotoxicol. Environ. Saf.* **2015**, *120*, 400–408. [CrossRef] [PubMed]
66. Dakal, T.C.; Kumar, A.; Majumdar, R.S.; Yadav, V. Mechanistic basis of antimicrobial actions of silver nanoparticles. *Front. Microbiol.* **2016**, *7*, 1831. [CrossRef] [PubMed]
67. Majeed, S.; Danish, M.; Binti Zahrudin, A.H.; Dash, G.K. Biosynthesis and characterization of silver nanoparticles from fungal species and its antibacterial and anticancer effect. *Karbala Int. J. Mod. Sci.* **2018**, *4*, 86–92. [CrossRef]
68. Saravanan, M.; Arokiyaraj, S.; Lakshmi, T.; Pugazhendhi, A. Synthesis of silver nanoparticles from phenerochaete chrysosporium (MTCC-787) and their antibacterial activity against human pathogenic bacteria. *Microb. Pathog.* **2018**, *117*, 68–72. [CrossRef] [PubMed]
69. Dastafkan, K.; Khajeh, M.; Bohlooli, M.; Ghaffari-Moghaddam, M.; Sheibani, N. Mechanism and behavior of silver nanoparticles in aqueous medium as adsorbent. *Talanta* **2015**, *144*, 1377–1386. [CrossRef] [PubMed]
70. Lim, Y.H.; Tiemann, K.M.; Heo, G.S.; Wagers, P.O.; Rezenom, Y.H.; Zhang, S.; Zhang, F.; Youngs, W.J.; Hunstad, D.A.; Wooley, K.L. Preparation and in vitro antimicrobial activity of silver-bearing degradable polymeric nanoparticles of polyphosphoester-block-poly(L-lactide). *ACS Nano* **2015**, *9*, 1995–2008. [CrossRef] [PubMed]
71. Schneider, G. Antimicrobial silver nanoparticles—Regulatory situation in the European Union. *Mater. Today Proc.* **2017**, *4*, S200–S207. [CrossRef]
72. Radzig, M.A.; Nadtochenko, V.A.; Koksharova, O.A.; Kiwi, J.; Lipasova, V.A.; Khmel, I.A. Antibacterial effects of silver nanoparticles on gram-negative bacteria: Influence on the growth and biofilms formation, mechanisms of action. *Colloids Surf. B Biointerfaces* **2013**, *102*, 300–306. [CrossRef] [PubMed]

73. Ribeiro, S.M.; Felicio, M.R.; Boas, E.V.; Goncalves, S.; Costa, F.F.; Samy, R.P.; Santos, N.C.; Franco, O.L. New frontiers for anti-biofilm drug development. *Pharmacol. Ther.* **2016**, *160*, 133–144. [CrossRef] [PubMed]
74. Barker, L.K.; Giska, J.R.; Radniecki, T.S.; Semprini, L. Effects of short- and long-term exposure of silver nanoparticles and silver ions to nitrosomonas europaea biofilms and planktonic cells. *Chemosphere* **2018**, *206*, 606–614. [CrossRef] [PubMed]
75. Joo, S.H.; Aggarwal, S. Factors impacting the interactions of engineered nanoparticles with bacterial cells and biofilms: Mechanistic insights and state of knowledge. *J. Environ. Manag.* **2018**, *225*, 62–74. [CrossRef] [PubMed]
76. Choi, Y.; Kim, H.-A.; Kim, K.-W.; Lee, B.-T. Comparative toxicity of silver nanoparticles and silver ions to *Escherichia coli*. *J. Environ. Sci.* **2018**, *66*, 50–60. [CrossRef] [PubMed]
77. Ramezanpour, M.; Leung, S.S.W.; Delgado-Magnero, K.H.; Bashe, B.Y.M.; Thewalt, J.; Tieleman, D.P. Computational and experimental approaches for investigating nanoparticle-based drug delivery systems. *Biochim. Biophys. Acta (BBA) Biomembr.* **2016**, *1858*, 1688–1709. [CrossRef] [PubMed]
78. Jahangirian, H.; Lemraski, E.G.; Webster, T.J.; Rafiee-Moghaddam, R.; Abdollahi, Y. A review of drug delivery systems based on nanotechnology and green chemistry: Green nanomedicine. *Int. J. Nanomed.* **2017**, *12*, 2957–2978. [CrossRef] [PubMed]
79. Jiang, Q.; Yu, S.; Li, X.; Ma, C.; Li, A. Evaluation of local anesthetic effects of lidocaine-ibuprofen ionic liquid stabilized silver nanoparticles in male swiss mice. *J. Photochem. Photobiol. B Biol.* **2018**, *178*, 367–370. [CrossRef] [PubMed]
80. Karthik, C.S.; Manukumar, H.M.; Ananda, A.P.; Nagashree, S.; Rakesh, K.P.; Mallesha, L.; Qin, H.-L.; Umesha, S.; Mallu, P.; Krishnamurthy, N.B. Synthesis of novel benzodioxane midst piperazine moiety decorated chitosan silver nanoparticle against biohazard pathogens and as potential anti-inflammatory candidate: A molecular docking studies. *Int. J. Biol. Macromol.* **2018**, *108*, 489–502. [CrossRef] [PubMed]
81. Soni, N.; Dhiman, R.C. Phytochemical, anti-oxidant, larvicidal, and antimicrobial activities of castor (*Ricinus communis*) synthesized silver nanoparticles. *Chin. Herb. Med.* **2017**, *9*, 289–294. [CrossRef]
82. Arumai Selvan, D.; Mahendiran, D.; Senthil Kumar, R.; Kalilur Rahiman, A. Garlic, green tea and turmeric extracts-mediated green synthesis of silver nanoparticles: Phytochemical, antioxidant and in vitro cytotoxicity studies. *J. Photochem. Photobiol. B Biol.* **2018**, *180*, 243–252. [CrossRef] [PubMed]
83. Al-Obaidi, H.; Kalgudi, R.; Zariwala, M.G. Fabrication of inhaled hybrid silver/ciprofloxacin nanoparticles with synergetic effect against pseudomonas aeruginosa. *Eur. J. Pharm. Biopharm.* **2018**, *128*, 27–35. [CrossRef] [PubMed]
84. Kaur, A.; Goyal, D.; Kumar, R. Surfactant mediated interaction of vancomycin with silver nanoparticles. *Appl. Surf. Sci.* **2018**, *449*, 23–30. [CrossRef]
85. Petrov, P.D.; Yoncheva, K.; Gancheva, V.; Konstantinov, S.; Trzebicka, B. Multifunctional block copolymer nanocarriers for co-delivery of silver nanoparticles and curcumin: Synthesis and enhanced efficacy against tumor cells. *Eur. Polym. J.* **2016**, *81*, 24–33. [CrossRef]
86. Tiwari, G.; Tiwari, R.; Sriwastawa, B.; Bhati, L.; Pandey, S.; Pandey, P.; Bannerjee, S.K. Drug delivery systems: An updated review. *Int. J. Pharm. Investig.* **2012**, *2*, 2–11. [CrossRef] [PubMed]
87. KJ, P. Multi-functional silver nanoparticles for drug delivery: A review. *Int. J. Curr. Pharm. Rev. Res.* **2017**, *9*, 1–5.
88. Tahseen, Q.A. Silver Nanoparticles as Drug Delivery Systems. Ph.D. Dissertations, Louisiana State University, Baton Rouge, LA, USA, 2013.
89. Anandhakumar, S.; Mahalakshmi, V.; Raichur, A.M. Silver nanoparticles modified nanocapsules for ultrasonically activated drug delivery. *Mater. Sci. Eng. C* **2012**, *32*, 2349–2355. [CrossRef]
90. Bagherzade, G.; Tavakoli, M.M.; Namaei, M.H. Green synthesis of silver nanoparticles using aqueous extract of saffron (*Crocus sativus* L.) wastages and its antibacterial activity against six bacteria. *Asian Pac. J. Trop. Biomed.* **2017**, *7*, 227–233. [CrossRef]
91. Khadka, P.; Ro, J.; Kim, H.; Kim, I.; Kim, J.T.; Kim, H.; Cho, J.M.; Yun, G.; Lee, J. Pharmaceutical particle technologies: An approach to improve drug solubility, dissolution and bioavailability. *Asian J. Pharm. Sci.* **2014**, *9*, 304–316. [CrossRef]
92. Kumar, B.; Jalodia, K.; Kumar, P.; Gautam, H.K. Recent advances in nanoparticle-mediated drug delivery. *J. Drug Deliv. Sci. Technol.* **2017**, *41*, 260–268. [CrossRef]
93. Venugopal, K.; Rather, H.A.; Rajagopal, K.; Shanthi, M.P.; Sheriff, K.; Illiyas, M.; Rather, R.A.; Manikandan, E.; Uvarajan, S.; Bhaskar, M.; et al. Synthesis of silver nanoparticles (Ag NPs) for anticancer activities

(MCF 7 breast and A549 lung cell lines) of the crude extract of *Syzygium* aromaticum. *J. Photochem. Photobiol. B Biol.* **2017**, *167*, 282–289. [CrossRef] [PubMed]
94. Benyettou, F.; Rezgui, R.; Ravaux, F.; Jaber, T.; Blumer, K.; Jouiad, M.; Motte, L.; Olsen, J.C.; Platas-Iglesias, C.; Magzoub, M.; et al. Synthesis of silver nanoparticles for the dual delivery of doxorubicin and alendronate to cancer cells. *J. Mater. Chem. B* **2015**, *3*, 7237–7245. [CrossRef]
95. Barbinta-Patrascu, M.E.; Badea, N.; Pirvu, C.; Bacalum, M.; Ungureanu, C.; Nadejde, P.L.; Ion, C.; Rau, I. Multifunctional soft hybrid bio-platforms based on nano-silver and natural compounds. *Mater. Sci. Eng. C* **2016**, *69*, 922–932. [CrossRef] [PubMed]
96. Patra, S.; Mukherjee, S.; Barui, A.K.; Ganguly, A.; Sreedhar, B.; Patra, C.R. Green synthesis, characterization of gold and silver nanoparticles and their potential application for cancer therapeutics. *Mater. Sci. Eng. C* **2015**, *53*, 298–309. [CrossRef] [PubMed]
97. Ding, Q.; Liu, D.; Guo, D.; Yang, F.; Pang, X.; Che, R.; Zhou, N.; Xie, J.; Sun, J.; Huang, Z.; et al. Shape-controlled fabrication of magnetite silver hybrid nanoparticles with high performance magnetic hyperthermia. *Biomaterials* **2017**, *124*, 35–46. [CrossRef] [PubMed]
98. Poudel, B.K.; Soe, Z.C.; Ruttala, H.B.; Gupta, B.; Ramasamy, T.; Thapa, R.K.; Gautam, M.; Ou, W.; Nguyen, H.T.; Jeong, J.-H.; et al. In situ fabrication of mesoporous silica-coated silver-gold hollow nanoshell for remotely controllable chemo-photothermal therapy via phase-change molecule as gatekeepers. *Int. J. Pharm.* **2018**, *548*, 92–103. [CrossRef] [PubMed]
99. Díaz-Cruz, C.; Alonso Nuñez, G.; Espinoza-Gómez, H.; Flores-López, L.Z. Effect of molecular weight of peg or pva as reducing-stabilizing agent in the green synthesis of silver-nanoparticles. *Eur. Polym. J.* **2016**, *83*, 265–277. [CrossRef]
100. Hefni, H.H.H.; Azzam, E.M.; Badr, E.A.; Hussein, M.; Tawfik, S.M. Synthesis, characterization and anticorrosion potentials of chitosan-g-peg assembled on silver nanoparticles. *Int. J. Biol. Macromol.* **2016**, *83*, 297–305. [CrossRef] [PubMed]
101. Yang, H.; Chen, T.; Wang, H.; Bai, S.; Guo, X. One-pot rapid synthesis of high aspect ratio silver nanowires for transparent conductive electrodes. *Mater. Res. Bull.* **2018**, *102*, 79–85. [CrossRef]
102. Gao, H.; Yang, H.; Wang, C. Controllable preparation and mechanism of nano-silver mediated by the microemulsion system of the clove oil. *Results Phys.* **2017**, *7*, 3130–3136. [CrossRef]
103. Rivera-Rangel, R.D.; González-Muñoz, M.P.; Avila-Rodriguez, M.; Razo-Lazcano, T.A.; Solans, C. Green synthesis of silver nanoparticles in oil-in-water microemulsion and nano-emulsion using geranium leaf aqueous extract as a reducing agent. *Colloids Surf. A Physicochem. Eng. Asp.* **2018**, *536*, 60–67. [CrossRef]
104. Clemente, A.; Moreno, N.; Lobera, M.P.; Balas, F.; Santamaria, J. Versatile hollow fluorescent metal-silica nanohybrids through a modified microemulsion synthesis route. *J. Colloid Interface Sci.* **2018**, *513*, 497–504. [CrossRef] [PubMed]
105. Hanh, T.T.; Thu, N.T.; Quoc, L.A.; Hien, N.Q. Synthesis and characterization of silver/diatomite nanocomposite by electron beam irradiation. *Radiat. Phys. Chem.* **2017**, *139*, 141–146. [CrossRef]
106. Dhayagude, A.C.; Das, A.; Joshi, S.S.; Kapoor, S. γ-radiation induced synthesis of silver nanoparticles in aqueous poly (N-vinylpyrrolidone) solution. *Colloids Surf. A Physicochem. Eng. Asp.* **2018**, *556*, 148–156. [CrossRef]
107. Zaheer, Z.; Aazam, E.S. Cetyltrimethylammonium bromide assisted synthesis of silver nanoparticles and their catalytic activity. *J. Mol. Liq.* **2017**, *242*, 1035–1041. [CrossRef]
108. Lopes, C.R.B.; Courrol, L.C. Green synthesis of silver nanoparticles with extract of mimusops coriacea and light. *J. Lumin.* **2018**, *199*, 183–187. [CrossRef]
109. Rai, M.; Ingle, A.P.; Gupta, I.; Brandelli, A. Bioactivity of noble metal nanoparticles decorated with biopolymers and their application in drug delivery. *Int. J. Pharm.* **2015**, *496*, 159–172. [CrossRef] [PubMed]
110. Sarkar, S.; Das, R. Shape effect on the optical properties of anisotropic silver nanocrystals. *J. Lumin.* **2018**, *198*, 464–470. [CrossRef]
111. Delgado-Beleño, Y.; Martinez-Nuñez, C.E.; Cortez-Valadez, M.; Flores-López, N.S.; Flores-Acosta, M. Optical properties of silver, silver sulfide and silver selenide nanoparticles and antibacterial applications. *Mater. Res. Bull.* **2018**, *99*, 385–392. [CrossRef]
112. Dos Santos Courrol, D.; Regina Borges Lopes, C.; da Silva Cordeiro, T.; Regina Franzolin, M.; Dias Vieira Junior, N.; Elgul Samad, R.; Coronato Courrol, L. Optical properties and antimicrobial effects of silver

nanoparticles synthesized by femtosecond laser photoreduction. *Opt. Laser Technol.* **2018**, *103*, 233–238. [CrossRef]

113. Brown, P.K.; Qureshi, A.T.; Moll, A.N.; Hayes, D.J.; Monroe, W.T. Silver nanoscale antisense drug delivery system for photoactivated gene silencing. *ACS Nano* **2013**, *7*, 2948–2959. [CrossRef] [PubMed]
114. Heilman, S.; Silva, L.G.A. Silver and titanium nanoparticles used as coating on polyurethane catheters. *J. Nano Res.* **2017**, *47*, 17–23. [CrossRef]
115. Thomas, R.; Mathew, S.; Nayana, A.R.; Mathews, J.; Radhakrishnan, E.K. Microbially and phytofabricated agnps with different mode of bactericidal action were identified to have comparable potential for surface fabrication of central venous catheters to combat staphylococcus aureus biofilm. *J. Photochem. Photobiol. B Biol.* **2017**, *171*, 96–103. [CrossRef] [PubMed]
116. Wu, K.; Yang, Y.; Zhang, Y.; Deng, J.; Lin, C. Antimicrobial activity and cytocompatibility of silver nanoparticles coated catheters via a biomimetic surface functionalization strategy. *Int. J. Nanomed.* **2015**, *10*, 7241–7252.
117. Roe, D.; Karandikar, B.; Bonn-Savage, N.; Gibbins, B.; Roullet, J.B. Antimicrobial surface functionalization of plastic catheters by silver nanoparticles. *J. Antimicrob. Chemother.* **2008**, *61*, 869–876. [CrossRef] [PubMed]
118. Kumar, C.G.; Sujitha, P. Green synthesis of kocuran-functionalized silver glyconanoparticles for use as antibiofilm coatings on silicone urethral catheters. *Nanotechnology* **2014**, *25*, 325101. [CrossRef] [PubMed]
119. Rtimi, S.; Sanjines, R.; Pulgarin, C.; Kiwi, J. Microstructure of cu–ag uniform nanoparticulate films on polyurethane 3D catheters: Surface properties. *ACS Appl. Mater. Interfaces* **2016**, *8*, 56–63. [CrossRef] [PubMed]
120. Ballottin, D.; Fulaz, S.; Cabrini, F.; Tsukamoto, J.; Durán, N.; Alves, O.L.; Tasic, L. Antimicrobial textiles: Biogenic silver nanoparticles against candida and xanthomonas. *Mater. Sci. Eng. C* **2017**, *75*, 582–589. [CrossRef] [PubMed]
121. Su, C.-H.; Kumar, G.V.; Adhikary, S.; Velusamy, P.; Pandian, K.; Anbu, P. Preparation of cotton fabric using sodium alginate-coated nanoparticles to protect against nosocomial pathogens. *Biochem. Eng. J.* **2017**, *117*, 28–35. [CrossRef]
122. Zhang, M.; Lin, H.; Wang, Y.; Yang, G.; Zhao, H.; Sun, D. Fabrication and durable antibacterial properties of 3D porous wet electrospun rcsc/pcl nanofibrous scaffold with silver nanoparticles. *Appl. Surf. Sci.* **2017**, *414*, 52–62. [CrossRef]
123. Alippilakkotte, S.; Kumar, S.; Sreejith, L. Fabrication of pla/ag nanofibers by green synthesis method using momordica charantia fruit extract for wound dressing applications. *Colloids Surf. A Physicochem. Eng. Asp.* **2017**, *529*, 771–782. [CrossRef]
124. Li, R.; He, M.; Li, T.; Zhang, L. Preparation and properties of cellulose/silver nanocomposite fibers. *Carbohydr. Polym.* **2015**, *115*, 269–275. [CrossRef] [PubMed]
125. Biswas, P.; Bandyopadhyaya, R. Biofouling prevention using silver nanoparticle impregnated polyethersulfone (PES) membrane: *E. coli* cell-killing in a continuous cross-flow membrane module. *J. Colloid Interface Sci.* **2017**, *491*, 13–26. [CrossRef] [PubMed]
126. Benavente, J.; García, M.E.; Urbano, N.; López-Romero, J.M.; Contreras-Cáceres, R.C.; Casado-Rodríguez, M.A.; Moscoso, A.; Hierrezuelo, J. Inclusion of silver nanoparticles for improving regenerated cellulose membrane performance and reduction of biofouling. *Int. J. Biol. Macromol.* **2017**, *103*, 758–763. [CrossRef] [PubMed]
127. Štular, D.; Jerman, I.; Naglič, I.; Simončič, B.; Tomšič, B. Embedment of silver into temperature- and ph-responsive microgel for the development of smart textiles with simultaneous moisture management and controlled antimicrobial activities. *Carbohydr. Polym.* **2017**, *159*, 161–170. [CrossRef] [PubMed]
128. Ding, L.; Shan, X.; Zhao, X.; Zha, H.; Chen, X.; Wang, J.; Cai, C.; Wang, X.; Li, G.; Hao, J.; et al. Spongy bilayer dressing composed of chitosan–Ag nanoparticles and chitosan–*Bletilla striata* polysaccharide for wound healing applications. *Carbohydr. Polym.* **2017**, *157*, 1538–1547. [CrossRef] [PubMed]

129. Stevens, K.N.; Crespo-Biel, O.; van den Bosch, E.E.; Dias, A.A.; Knetsch, M.L.; Aldenhoff, Y.B.; van der Veen, F.H.; Maessen, J.G.; Stobberingh, E.E.; Koole, L.H. The relationship between the antimicrobial effect of catheter coatings containing silver nanoparticles and the coagulation of contacting blood. *Biomaterials* **2009**, *30*, 3682–3690. [CrossRef] [PubMed]
130. Fufa, O.; Andronescu, E.; Grumezescu, V.; Holban, A.M.; Mogoanta, L.; Mogosanu, G.D.; Socol, G.; Iordache, F.; Chifiriuc, M.C.; Grumezescu, A.M. Silver nanostructurated surfaces prepared by maple for biofilm prevention. *Biointerface Res. Appl. Chem.* **2015**, *5*, 1011–1017.
131. Mala, R.; Annie Aglin, A.; Ruby Celsia, A.S.; Geerthika, S.; Kiruthika, N.; VazagaPriya, C.; Srinivasa Kumar, K. Foley catheters functionalised with a synergistic combination of antibiotics and silver nanoparticles resist biofilm formation. *IET Nanobiotechnol.* **2017**, *11*, 612–620. [CrossRef] [PubMed]
132. Jishma, P.; Narayanan, R.; Snigdha, S.; Thomas, R.; Radhakrishnan, E.K. Rapid degradative effect of microbially synthesized silver nanoparticles on textile dye in presence of sunlight. *Biocatal. Agric. Biotechnol.* **2018**, *14*, 410–417. [CrossRef]
133. Antonelli, M.; De Pascale, G.; Ranieri, V.M.; Pelaia, P.; Tufano, R.; Piazza, O.; Zangrillo, A.; Ferrario, A.; De Gaetano, A.; Guaglianone, E.; et al. Comparison of triple-lumen central venous catheters impregnated with silver nanoparticles (AgTive®) vs. conventional catheters in intensive care unit patients. *J. Hosp. Infect.* **2012**, *82*, 101–107. [CrossRef] [PubMed]
134. Stevens, K.N.J.; Croes, S.; Boersma, R.S.; Stobberingh, E.E.; van der Marel, C.; van der Veen, F.H.; Knetsch, M.L.W.; Koole, L.H. Hydrophilic surface coatings with embedded biocidal silver nanoparticles and sodium heparin for central venous catheters. *Biomaterials* **2011**, *32*, 1264–1269. [CrossRef] [PubMed]
135. Pollini, M.; Paladini, F.; Catalano, M.; Taurino, A.; Licciulli, A.; Maffezzoli, A.; Sannino, A. Antibacterial coatings on haemodialysis catheters by photochemical deposition of silver nanoparticles. *J. Mater. Sci. Mater. Med.* **2011**, *22*, 2005–2012. [CrossRef] [PubMed]
136. Aflori, M. Surface characterization of peritoneal dialysis catheter containing silver nanoparticles. *Rev. Roum. Chim.* **2014**, *59*, 523–526.
137. Ballo, M.K.; Rtimi, S.; Pulgarin, C.; Hopf, N.; Berthet, A.; Kiwi, J.; Moreillon, P.; Entenza, J.M.; Bizzini, A. In vitro and in vivo effectiveness of an innovative silver-copper nanoparticle coating of catheters to prevent methicillin-resistant staphylococcus aureus infection. *Antimicrob. Agents Chemother.* **2016**, *60*, 5349–5356. [CrossRef] [PubMed]
138. Dayyoub, E.; Frant, M.; Pinnapireddy, S.R.; Liefeith, K.; Bakowsky, U. Antibacterial and anti-encrustation biodegradable polymer coating for urinary catheter. *Int. J. Pharm.* **2017**, *531*, 205–214. [CrossRef] [PubMed]
139. Samuel, U.; Guggenbichler, J.P. Prevention of catheter-related infections: The potential of a new nano-silver impregnated catheter. *Int. J. Antimicrob. Agents* **2004**, *23* (Suppl. 1), S75–S78. [CrossRef]
140. Cheng, L.; Zhang, K.; Weir, M.D.; Melo, M.A.; Zhou, X.; Xu, H.H. Nanotechnology strategies for antibacterial and remineralizing composites and adhesives to tackle dental caries. *Nanomedicine* **2015**, *10*, 627–641. [CrossRef] [PubMed]
141. Lee, D.; Lee, S.J.; Moon, J.-H.; Kim, J.H.; Heo, D.N.; Bang, J.B.; Lim, H.-N.; Kwon, I.K. Preparation of antibacterial chitosan membranes containing silver nanoparticles for dental barrier membrane applications. *J. Ind. Eng. Chem.* **2018**. [CrossRef]
142. Divakar, D.D.; Jastaniyah, N.T.; Altamimi, H.G.; Alnakhli, Y.O.; Muzaheed; Alkheraif, A.A.; Haleem, S. Enhanced antimicrobial activity of naturally derived bioactive molecule chitosan conjugated silver nanoparticle against dental implant pathogens. *Int. J. Biol. Macromol.* **2018**, *108*, 790–797. [CrossRef] [PubMed]
143. Kaur, P.; Luthra, R. Silver nanoparticles in dentistry: An emerging trend. *SRM J. Res. Dent. Sci.* **2016**, *7*, 162–165. [CrossRef]
144. Bapat, R.A.; Chaubal, T.V.; Joshi, C.P.; Bapat, P.R.; Choudhury, H.; Pandey, M.; Gorain, B.; Kesharwani, P. An overview of application of silver nanoparticles for biomaterials in dentistry. *Mater. Sci. Eng. C* **2018**, *91*, 881–898. [CrossRef] [PubMed]
145. Vogel, K.; Westphal, N.; Salz, D.; Thiel, K.; Wittig, L.; Ciacchi, L.C.; Grunwald, I. Dental implants coated with a durable and antibacterial film. *Surf. Innov.* **2015**, *3*, 27–38. [CrossRef]
146. Noronha, V.T.; Paula, A.J.; Durán, G.; Galembeck, A.; Cogo-Müller, K.; Franz-Montan, M.; Durán, N. Silver nanoparticles in dentistry. *Dent. Mater.* **2017**, *33*, 1110–1126. [CrossRef] [PubMed]

147. Correa, J.M.; Mori, M.; Sanches, H.L.; da Cruz, A.D.; Poiate, E., Jr.; Poiate, I.A. Silver nanoparticles in dental biomaterials. *Int. J. Biomater.* **2015**, *2015*, 485275. [CrossRef] [PubMed]
148. Manikandan, V.; Velmurugan, P.; Park, J.H.; Chang, W.S.; Park, Y.J.; Jayanthi, P.; Cho, M.; Oh, B.T. Green synthesis of silver oxide nanoparticles and its antibacterial activity against dental pathogens. *3 Biotech* **2017**, *7*, 72. [CrossRef] [PubMed]
149. Priyadarsini, S.; Mukherjee, S.; Mishra, M. Nanoparticles used in dentistry: A review. *J. Oral Biol. Craniofac. Res.* **2018**, *8*, 58–67. [CrossRef] [PubMed]
150. Zhang, N.; Melo, M.A.S.; Antonucci, J.M.; Lin, N.J.; Lin-Gibson, S.; Bai, Y.; Xu, H.H.K. Novel dental cement to combat biofilms and reduce acids for orthodontic applications to avoid enamel demineralization. *Materials* **2016**, *9*, 413. [CrossRef] [PubMed]
151. Nguyen, S.; Hiorth, M. Advanced drug delivery systems for local treatment of the oral cavity. *Ther. Deliv.* **2015**, *6*, 595–608. [CrossRef] [PubMed]
152. Elias Santos, V.; Targino, A.; Pelagio Flores, M.A.; de Luna Freire Pessoa, H.; Galembeck, A.; Rosenblatt, A. Antimicrobial activity of silver nanoparticles in treating dental caries. *RFO* **2014**, *18*, 312–315. [CrossRef]
153. Chambers, C.; Stewart, S.B.; Su, B.; Jenkinson, H.F.; Sandy, J.R.; Ireland, A.J. Silver doped titanium dioxide nanoparticles as antimicrobial additives to dental polymers. *Dent. Mater.* **2017**, *33*, e115–e123. [CrossRef] [PubMed]
154. Mahross, H.Z.; Baroudi, K. Effect of silver nanoparticles incorporation on viscoelastic properties of acrylic resin denture base material. *Eur. J. Dent.* **2015**, *9*, 207–212. [PubMed]
155. Pokrowiecki, R.; Zareba, T.; Szaraniec, B.; Palka, K.; Mielczarek, A.; Menaszek, E.; Tyski, S. In vitro studies of nanosilver-doped titanium implants for oral and maxillofacial surgery. *Int. J. Nanomed.* **2017**, *12*, 4285–4297. [CrossRef] [PubMed]
156. Lee, S.J.; Heo, M.; Lee, D.; Han, S.; Moon, J.-H.; Lim, H.-N.; Kwon, I.K. Preparation and characterization of antibacterial orthodontic resin containing silver nanoparticles. *Appl. Surf. Sci.* **2018**, *432*, 317–323. [CrossRef]
157. Ai, M.; Du, Z.; Zhu, S.; Geng, H.; Zhang, X.; Cai, Q.; Yang, X. Composite resin reinforced with silver nanoparticles–laden hydroxyapatite nanowires for dental application. *Dent. Mater.* **2017**, *33*, 12–22. [CrossRef] [PubMed]
158. Paiva, L.; Fidalgo, T.K.S.; da Costa, L.P.; Maia, L.C.; Balan, L.; Anselme, K.; Ploux, L.; Thiré, R.M.S.M. Antibacterial properties and compressive strength of new one-step preparation silver nanoparticles in glass ionomer cements (NanoAg-GIC). *J. Dent.* **2018**, *69*, 102–109. [CrossRef] [PubMed]
159. Slane, J.; Vivanco, J.; Rose, W.; Ploeg, H.-L.; Squire, M. Mechanical, material, and antimicrobial properties of acrylic bone cement impregnated with silver nanoparticles. *Mater. Sci. Eng. C* **2015**, *48*, 188–196. [CrossRef] [PubMed]
160. Freire, P.L.L.; Albuquerque, A.J.R.; Farias, I.A.P.; da Silva, T.G.; Aguiar, J.S.; Galembeck, A.; Flores, M.A.P.; Sampaio, F.C.; Stamford, T.C.M.; Rosenblatt, A. Antimicrobial and cytotoxicity evaluation of colloidal chitosan—Silver nanoparticles—Fluoride nanocomposites. *Int. J. Biol. Macromol.* **2016**, *93*, 896–903. [CrossRef] [PubMed]
161. Wang, J.; Dong, X.; Yu, Q.; Baker, S.N.; Li, H.; Larm, N.E.; Baker, G.A.; Chen, L.; Tan, J.; Chen, M. Incorporation of antibacterial agent derived deep eutectic solvent into an active dental composite. *Dent. Mater.* **2017**, *33*, 1445–1455. [CrossRef] [PubMed]
162. Natale, L.C.; Alania, Y.; Rodrigues, M.C.; Simões, A.; de Souza, D.N.; de Lima, E.; Arana-Chavez, V.E.; Hewer, T.L.R.; Hiers, R.; Esteban-Florez, F.L.; et al. Synthesis and characterization of silver phosphate/calcium phosphate mixed particles capable of silver nanoparticle formation by photoreduction. *Mater. Sci. Eng. C* **2017**, *76*, 464–471. [CrossRef] [PubMed]
163. Gunputh, U.F.; Le, H.; Handy, R.D.; Tredwin, C. Anodised TiO_2 nanotubes as a scaffold for antibacterial silver nanoparticles on titanium implants. *Mater. Sci. Eng. C* **2018**, *91*, 638–644. [CrossRef] [PubMed]
164. Ferraris, S.; Spriano, S.; Miola, M.; Bertone, E.; Allizond, V.; Cuffini, A.M.; Banche, G. Surface modification of titanium surfaces through a modified oxide layer and embedded silver nanoparticles: Effect of reducing/stabilizing agents on precipitation and properties of the nanoparticles. *Surf. Coat. Technol.* **2018**, *344*, 177–189. [CrossRef]
165. Garcia-Contreras, R.; Argueta-Figueroa, L.; Mejia-Rubalcava, C.; Jimenez-Martinez, R.; Cuevas-Guajardo, S.; Sanchez-Reyna, P.A.; Mendieta-Zeron, H. Perspectives for the use of silver nanoparticles in dental practice. *Int. Dent. J.* **2011**, *61*, 297–301. [CrossRef] [PubMed]

166. Wilkinson, L.J.; White, R.J.; Chipman, J.K. Silver and nanoparticles of silver in wound dressings: A review of efficacy and safety. *J. Wound Care* **2011**, *20*, 543–549. [CrossRef] [PubMed]
167. Chowdhury, S.; De, M.; Guha, R.; Batabyal, S.; Samanta, I.; Hazra Samir, K.; Ghosh Tamal, K.; Konar, A.; Hazra, S. Influence of silver nanoparticles on post-surgical wound healing following topical application. *Eur. J. Nanomed.* **2014**, *6*, 237. [CrossRef]
168. You, C.; Li, Q.; Wang, X.; Wu, P.; Ho, J.K.; Jin, R.; Zhang, L.; Shao, H.; Han, C. Silver nanoparticle loaded collagen/chitosan scaffolds promote wound healing via regulating fibroblast migration and macrophage activation. *Sci. Rep.* **2017**, *7*, 10489. [CrossRef] [PubMed]
169. Zulkifli, F.H.; Hussain, F.S.J.; Zeyohannes, S.S.; Rasad, M.S.B.A.; Yusuff, M.M. A facile synthesis method of hydroxyethyl cellulose-silver nanoparticle scaffolds for skin tissue engineering applications. *Mater. Sci. Eng. C* **2017**, *79*, 151–160. [CrossRef] [PubMed]
170. Gong, C.P.; Li, S.C.; Wang, R.Y. Development of biosynthesized silver nanoparticles based formulation for treating wounds during nursing care in hospitals. *J. Photochem. Photobiol. B Biol.* **2018**, *183*, 137–141. [CrossRef] [PubMed]
171. Hendi, A. Silver nanoparticles mediate differential responses in some of liver and kidney functions during skin wound healing. *J. King Saud Univ. Sci.* **2011**, *23*, 47–52. [CrossRef]
172. Hebeish, A.; El-Rafie, M.H.; El-Sheikh, M.A.; Seleem, A.A.; El-Naggar, M.E. Antimicrobial wound dressing and anti-inflammatory efficacy of silver nanoparticles. *Int. J. Biol. Macromol.* **2014**, *65*, 509–515. [CrossRef] [PubMed]
173. Konop, M.; Damps, T.; Misicka, A.; Rudnicka, L. Certain aspects of silver and silver nanoparticles in wound care: A minireview. *J. Nanomater.* **2016**, *2016*, 7614753. [CrossRef]
174. Yang, Y.; Hu, H. A review on antimicrobial silver absorbent wound dressings applied to exuding wounds. *J. Microb. Biochem. Technol.* **2015**, *7*, 228–233.
175. Larese, F.F.; D'Agostin, F.; Crosera, M.; Adami, G.; Renzi, N.; Bovenzi, M.; Maina, G. Human skin penetration of silver nanoparticles through intact and damaged skin. *Toxicology* **2009**, *255*, 33–37. [CrossRef] [PubMed]
176. Abdelgawad, A.M.; Hudson, S.M.; Rojas, O.J. Antimicrobial wound dressing nanofiber mats from multicomponent (chitosan/silver-NPs/polyvinyl alcohol) systems. *Carbohydr. Polym.* **2014**, *100*, 166–178. [CrossRef] [PubMed]
177. Cardoso, V.S.; de Carvalho Filgueiras, M.; Dutra, Y.M.; Teles, R.H.G.; de Araujo, A.R.; Primo, F.L.; Mafud, A.C.; Batista, L.F.; Mascarenhas, Y.P.; Paino, I.M.M.; et al. Collagen-based silver nanoparticles: Study on cell viability, skin permeation, and swelling inhibition. *Mater. Sci. Eng. C Mater. Biol. Appl.* **2017**, *74*, 382–388. [CrossRef] [PubMed]
178. Kumar, S.S.D.; Rajendran, N.K.; Houreld, N.N.; Abrahamse, H. Recent advances on silver nanoparticle and biopolymer-based biomaterials for wound healing applications. *Int. J. Biol. Macromol.* **2018**, *115*, 165–175. [CrossRef] [PubMed]
179. Bozaci, E.; Akar, E.; Ozdogan, E.; Demir, A.; Altinisik, A.; Seki, Y. Application of carboxymethylcellulose hydrogel based silver nanocomposites on cotton fabrics for antibacterial property. *Carbohydr. Polym.* **2015**, *134*, 128–135. [CrossRef] [PubMed]
180. Emam, H.E.; Saleh, N.H.; Nagy, K.S.; Zahran, M.K. Functionalization of medical cotton by direct incorporation of silver nanoparticles. *Int. J. Biol. Macromol.* **2015**, *78*, 249–256. [CrossRef] [PubMed]
181. Li, Z.; Wang, L.; Chen, S.; Feng, C.; Chen, S.; Yin, N.; Yang, J.; Wang, H.; Xu, Y. Facilely green synthesis of silver nanoparticles into bacterial cellulose. *Cellulose* **2015**, *22*, 373–383. [CrossRef]
182. Shao, W.; Liu, H.; Liu, X.; Sun, H.; Wang, S.; Zhang, R. Ph-responsive release behavior and anti-bacterial activity of bacterial cellulose-silver nanocomposites. *Int. J. Biol. Macromol.* **2015**, *76*, 209–217. [CrossRef] [PubMed]
183. Martins, A.F.; Monteiro, J.P.; Bonafé, E.G.; Gerola, A.P.; Silva, C.T.P.; Girotto, E.M.; Rubira, A.F.; Muniz, E.C. Bactericidal activity of hydrogel beads based on *N,N,N*-trimethyl chitosan/alginate complexes loaded with silver nanoparticles. *Chin. Chem. Lett.* **2015**, *26*, 1129–1132. [CrossRef]
184. Eghbalifam, N.; Frounchi, M.; Dadbin, S. Antibacterial silver nanoparticles in polyvinyl alcohol/sodium alginate blend produced by gamma irradiation. *Int. J. Biol. Macromol.* **2015**, *80*, 170–176. [CrossRef] [PubMed]
185. Zhao, X.; Li, Q.; Ma, X.; Quan, F.; Wang, J.; Xia, Y. The preparation of alginate–AgNPs composite fiber with green approach and its antibacterial activity. *J. Ind. Eng. Chem.* **2015**, *24*, 188–195. [CrossRef]

186. Mishra, M.; Kumar, H.; Tripathi, K. Diabetic delayed wound healing and the role of silver nanoparticles. *Dig. J. Nanomater. Bios.* **2008**, *3*, 49–54.
187. Rigo, C.; Ferroni, L.; Tocco, I.; Roman, M.; Munivrana, I.; Gardin, C.; Cairns, W.R.L.; Vindigni, V.; Azzena, B.; Barbante, C.; et al. Active silver nanoparticles for wound healing. *Int. J. Mol. Sci.* **2013**, *14*, 4817–4840. [CrossRef] [PubMed]
188. Correia, T.R.; Figueira, D.R.; de Sá, K.D.; Miguel, S.P.; Fradique, R.G.; Mendonça, A.G.; Correia, I.J. 3D printed scaffolds with bactericidal activity aimed for bone tissue regeneration. *Int. J. Biol. Macromol.* **2016**, *93*, 1432–1445. [CrossRef] [PubMed]
189. Castiglioni, S.; Cazzaniga, A.; Locatelli, L.; Maier, J.A.M. Silver nanoparticles in orthopedic applications: New insights on their effects on osteogenic cells. *Nanomaterials* **2017**, *7*, 124. [CrossRef] [PubMed]
190. Ralston, S.H. Bone structure and metabolism. *Medicine* **2013**, *41*, 581–585. [CrossRef]
191. Zhang, Y.; Zhai, D.; Xu, M.; Yao, Q.; Zhu, H.; Chang, J.; Wu, C. 3D-printed bioceramic scaffolds with antibacterial and osteogenic activity. *Biofabrication* **2017**, *9*, 025037. [CrossRef] [PubMed]
192. Aurore, V.; Caldana, F.; Blanchard, M.; Kharoubi Hess, S.; Lannes, N.; Mantel, P.Y.; Filgueira, L.; Walch, M. Silver-nanoparticles increase bactericidal activity and radical oxygen responses against bacterial pathogens in human osteoclasts. *Nanomed. Nanotechnol. Biol. Med.* **2018**, *14*, 601–607. [CrossRef] [PubMed]
193. Qing, T.; Mahmood, M.; Zheng, Y.; Biris, A.S.; Shi, L.; Casciano, D.A. A genomic characterization of the influence of silver nanoparticles on bone differentiation in MC3T3-E1 cells. *J. Appl. Toxicol. JAT* **2018**, *38*, 172–179. [CrossRef] [PubMed]
194. Brennan, S.A.; Ni Fhoghlu, C.; Devitt, B.M.; O'Mahony, F.J.; Brabazon, D.; Walsh, A. Silver nanoparticles and their orthopaedic applications. *Bone Jt. J.* **2015**, *97-B*, 582–589. [CrossRef] [PubMed]
195. Lu, H.; Liu, Y.; Guo, J.; Wu, H.; Wang, J.; Wu, G. Biomaterials with antibacterial and osteoinductive properties to repair infected bone defects. *Int. J. Mol. Sci.* **2016**, *17*, 334. [CrossRef] [PubMed]
196. Bharti, A.; Singh, S.; Meena, V.K.; Goyal, N. Structural characterization of silver-hydroxyapatite nanocomposite: A bone repair biomaterial. *Mater. Today Proc.* **2016**, *3*, 2113–2120. [CrossRef]
197. Lim, P.N.; Chang, L.; Thian, E.S. Development of nanosized silver-substituted apatite for biomedical applications: A review. *Nanomed. Nanotechnol. Biol. Med.* **2015**, *11*, 1331–1344. [CrossRef] [PubMed]
198. Lazić, V.; Smičiklas, I.; Marković, J.; Lončarević, D.; Dostanić, J.; Ahrenkiel, S.P.; Nedeljković, J.M. Antibacterial ability of supported silver nanoparticles by functionalized hydroxyapatite with 5-aminosalicylic acid. *Vacuum* **2018**, *148*, 62–68. [CrossRef]
199. Fu, C.; Zhang, X.; Savino, K.; Gabrys, P.; Gao, Y.; Chaimayo, W.; Miller, B.L.; Yates, M.Z. Antimicrobial silver-hydroxyapatite composite coatings through two-stage electrochemical synthesis. *Surf. Coat. Technol.* **2016**, *301*, 13–19. [CrossRef]
200. Andrade, F.A.C.; de Oliveira Vercik, L.C.; Monteiro, F.J.; da Silva Rigo, E.C. Preparation, characterization and antibacterial properties of silver nanoparticles–hydroxyapatite composites by a simple and eco-friendly method. *Ceram. Int.* **2016**, *42*, 2271–2280. [CrossRef]
201. Yu, W.-Z.; Zhang, Y.; Liu, X.; Xiang, Y.; Li, Z.; Wu, S. Synergistic antibacterial activity of multi components in lysozyme/chitosan/silver/hydroxyapatite hybrid coating. *Mater. Des.* **2018**, *139*, 351–362. [CrossRef]
202. Zhang, X.; Chaimayo, W.; Yang, C.; Yao, J.; Miller, B.L.; Yates, M.Z. Silver-hydroxyapatite composite coatings with enhanced antimicrobial activities through heat treatment. *Surf. Coat. Technol.* **2017**, *325*, 39–45. [CrossRef]
203. Geng, Z.; Wang, R.; Zhuo, X.; Li, Z.; Huang, Y.; Ma, L.; Cui, Z.; Zhu, S.; Liang, Y.; Liu, Y.; et al. Incorporation of silver and strontium in hydroxyapatite coating on titanium surface for enhanced antibacterial and biological properties. *Mater. Sci. Eng. C* **2017**, *71*, 852–861. [CrossRef] [PubMed]
204. Mirzaee, M.; Vaezi, M.; Palizdar, Y. Synthesis and characterization of silver doped hydroxyapatite nanocomposite coatings and evaluation of their antibacterial and corrosion resistance properties in simulated body fluid. *Mater. Sci. Eng. C* **2016**, *69*, 675–684. [CrossRef] [PubMed]
205. Zhou, K.; Dong, C.; Zhang, X.; Shi, L.; Chen, Z.; Xu, Y.; Cai, H. Preparation and characterization of nanosilver-doped porous hydroxyapatite scaffolds. *Ceram. Int.* **2015**, *41*, 1671–1676. [CrossRef]
206. Ciobanu, C.S.; Iconaru, S.L.; Pasuk, I.; Vasile, B.S.; Lupu, A.R.; Hermenean, A.; Dinischiotu, A.; Predoi, D. Structural properties of silver doped hydroxyapatite and their biocompatibility. *Mater. Sci. Eng. C* **2013**, *33*, 1395–1402. [CrossRef] [PubMed]

207. Ciobanu, G.; Ilisei, S.; Luca, C. Hydroxyapatite-silver nanoparticles coatings on porous polyurethane scaffold. *Mater. Sci. Eng. C* **2014**, *35*, 36–42. [CrossRef] [PubMed]
208. Jin, G.; Qin, H.; Cao, H.; Qian, S.; Zhao, Y.; Peng, X.; Zhang, X.; Liu, X.; Chu, P.K. Synergistic effects of dual zn/ag ion implantation in osteogenic activity and antibacterial ability of titanium. *Biomaterials* **2014**, *35*, 7699–7713. [CrossRef] [PubMed]
209. Hasan, A.; Waibhaw, G.; Saxena, V.; Pandey, L.M. Nano-biocomposite scaffolds of chitosan, carboxymethyl cellulose and silver nanoparticle modified cellulose nanowhiskers for bone tissue engineering applications. *Int. J. Biol. Macromol.* **2018**, *111*, 923–934. [CrossRef] [PubMed]
210. Strydom, S.J.; Rose, W.E.; Otto, D.P.; Liebenberg, W.; De Villiers, M.M. Poly(amidoamine) dendrimer-mediated synthesis and stabilization of silver sulfonamide nanoparticles with increased antibacterial activity. *Nanomedicine* **2013**, *9*, 85–93. [CrossRef] [PubMed]
211. González-Sánchez, M.I.; Perni, S.; Tommasi, G.; Morris, N.G.; Hawkins, K.; López-Cabarcos, E.; Prokopovich, P. Silver nanoparticle based antibacterial methacrylate hydrogels potential for bone graft applications. *Mater. Sci. Eng. C* **2015**, *50*, 332–340. [CrossRef] [PubMed]
212. Zhang, R.; Lee, P.; Lui, V.C.; Chen, Y.; Liu, X.; Lok, C.N.; To, M.; Yeung, K.W.; Wong, K.K. Silver nanoparticles promote osteogenesis of mesenchymal stem cells and improve bone fracture healing in osteogenesis mechanism mouse model. *Nanomed. Nanotechnol. Biol. Med.* **2015**, *11*, 1949–1959. [CrossRef] [PubMed]
213. Pauksch, L.; Hartmann, S.; Rohnke, M.; Szalay, G.; Alt, V.; Schnettler, R.; Lips, K. Biocompatibility of silver nanoparticles and silver ions in primary human mesenchymal stem cells and osteoblasts. *Acta Biomater.* **2013**, *10*, 439–449. [CrossRef] [PubMed]
214. De Sá, K.D.; Figueira, D.R.; Miguel, S.P.; Correia, T.R.; Silva, A.P.; Correia, I.J. 3D scaffolds coated with nanofibers displaying bactericidal activity for bone tissue applications. *Int. J. Polym. Mater. Polym. Biomater.* **2017**, *66*, 432–442. [CrossRef]
215. Thapa, R.K.; Kim, J.H.; Jeong, J.H.; Shin, B.S.; Choi, H.G.; Yong, C.S.; Kim, J.O. Silver nanoparticle-embedded graphene oxide-methotrexate for targeted cancer treatment. *Colloids Surf. B Biointerfaces* **2017**, *153*, 95–103. [CrossRef] [PubMed]
216. Rajeshkumar, S.; Malarkodi, C.; Vanaja, M.; Annadurai, G. Anticancer and enhanced antimicrobial activity of biosynthesizd silver nanoparticles against clinical pathogens. *J. Mol. Struct.* **2016**, *1116*, 165–173. [CrossRef]
217. Pongrac, I.M.; Ahmed, L.B.; Mlinarić, H.; Jurašin, D.D.; Pavičić, I.; Marjanović Čermak, A.M.; Milić, M.; Gajović, S.; Vinković Vrček, I. Surface coating affects uptake of silver nanoparticles in neural stem cells. *J. Trace Elem. Med. Biol.* **2017**. [CrossRef] [PubMed]
218. Mattea, F.; Vedelago, J.; Malano, F.; Gomez, C.; Strumia, M.; Valente, M. Silver nanoparticles in X-ray biomedical applications. *Radiat. Phys. Chem.* **2017**, *130*, 442–450. [CrossRef]
219. Vedelago, J.; Gomez, C.G.; Valente, M.; Mattea, F. Green synthesis of silver nanoparticles aimed at improving theranostics. *Radiat. Phys. Chem.* **2018**, *146*, 55–67. [CrossRef]
220. Sharma, H.; Mishra, P.K.; Talegaonkar, S.; Vaidya, B. Metal nanoparticles: A theranostic nanotool against cancer. *Drug Discov. Today* **2015**, *20*, 1143–1151. [CrossRef] [PubMed]
221. Jiang, W.; Rutherford, D.; Vuong, T.; Liu, H. Nanomaterials for treating cardiovascular diseases: A review. *Bioact. Mater.* **2017**, *2*, 185–198. [CrossRef] [PubMed]
222. Gonzalez, C.; Rosas-Hernandez, H.; Ramirez-Lee, M.A.; Salazar-Garcia, S.; Ali, S.F. Role of silver nanoparticles (AgNPs) on the cardiovascular system. *Arch. Toxicol.* **2016**, *90*, 493–511. [CrossRef] [PubMed]
223. Ramirez-Lee, M.A.; Aguirre-Banuelos, P.; Martinez-Cuevas, P.P.; Espinosa-Tanguma, R.; Chi-Ahumada, E.; Martinez-Castanon, G.A.; Gonzalez, C. Evaluation of cardiovascular responses to silver nanoparticles (AgNPs) in spontaneously hypertensive rats. *Nanomed. Nanotechnol. Biol. Med.* **2018**, *14*, 385–395. [CrossRef] [PubMed]
224. Ge, L.; Li, Q.; Wang, M.; Ouyang, J.; Li, X.; Xing, M.M.Q. Nanosilver particles in medical applications: Synthesis, performance, and toxicity. *Int. J. Nanomed.* **2014**, *9*, 2399–2407.
225. Rai, M.; Ingle, A.P.; Paralikar, P.; Gupta, I.; Medici, S.; Santos, C.A. Recent advances in use of silver nanoparticles as antimalarial agents. *Int. J. Pharm.* **2017**, *526*, 254–270. [CrossRef] [PubMed]
226. Chakrabarti, S.; Islam, J.; Hazarika, H.; Mazumder, B.; Raju, P.S.; Chattopadhyay, P. Safety profile of silver sulfadiazine-bfgf-loaded hydrogel for partial thickness burn wounds. *Cutan. Ocul. Toxicol.* **2018**, *37*, 258–266. [CrossRef] [PubMed]

227. Sung, A.Y.; Kim, T.-H. Physical properties of ophthalmic hydrogel polymer containing zinc oxide nanoparticles. *J. Chosun Nat. Sci.* **2013**, *6*, 76–81. [CrossRef]
228. Söderstjerna, E.; Bauer, P.; Cedervall, T.; Abdshill, H.; Johansson, F.; Johansson, U.E. Silver and gold nanoparticles exposure to in vitro cultured retina—Studies on nanoparticle internalization, apoptosis, oxidative stress, glial- and microglial activity. *PLoS ONE* **2014**, *9*, e105359. [CrossRef] [PubMed]
229. Weng, Y.; Liu, J.; Jin, S.; Guo, W.; Liang, X.; Hu, Z. Nanotechnology-based strategies for treatment of ocular disease. *Acta Pharm. Sin. B* **2017**, *7*, 281–291. [CrossRef] [PubMed]
230. Yee, W.; Selvaduray, G.; Hawkins, B. Characterization of silver nanoparticle-infused tissue adhesive for ophthalmic use. *J. Mech. Behav. Biomed. Mater.* **2016**, *55*, 67–74. [CrossRef] [PubMed]
231. Rizzello, L.; Pompa, P.P. Nanosilver-based antibacterial drugs and devices: Mechanisms, methodological drawbacks, and guidelines. *Chem. Soc. Rev.* **2014**, *43*, 1501–1518. [CrossRef] [PubMed]
232. Pinďáková, L.; Kašpárková, V.; Kejlová, K.; Dvořáková, M.; Krsek, D.; Jírová, D.; Kašparová, L. Behaviour of silver nanoparticles in simulated saliva and gastrointestinal fluids. *Int. J. Pharm.* **2017**, *527*, 12–20. [CrossRef] [PubMed]
233. Rai, M.; Kon, K.; Ingle, A.; Duran, N.; Galdiero, S.; Galdiero, M. Broad-spectrum bioactivities of silver nanoparticles: The emerging trends and future prospects. *Appl. Microbiol. Biotechnol.* **2014**, *98*, 1951–1961. [CrossRef] [PubMed]
234. Hadrup, N.; Lam, H.R. Oral toxicity of silver ions, silver nanoparticles and colloidal silver—A review. *Regul. Toxicol. Pharmacol.* **2014**, *68*, 1–7. [CrossRef] [PubMed]
235. Vazquez-Munoz, R.; Borrego, B.; Juarez-Moreno, K.; Garcia-Garcia, M.; Mota Morales, J.D.; Bogdanchikova, N.; Huerta-Saquero, A. Toxicity of silver nanoparticles in biological systems: Does the complexity of biological systems matter? *Toxicol. Lett.* **2017**, *276*, 11–20. [CrossRef] [PubMed]
236. Salarian, A.A.; Bahari, Y.; Hami, Z.; Soltani-Rezaee-Rad, M. Cephalexin nanoparticles: Synthesis, cytotoxicity and their synergistic antibacterial study in combination with silver nanoparticles. *Mater. Chem. Phys.* **2017**, *198*, 125–130. [CrossRef]
237. Stensberg, M.C.; Wei, Q.; McLamore, E.S.; Porterfield, D.M.; Wei, A.; Sepúlveda, M.S. Toxicological studies on silver nanoparticles: Challenges and opportunities in assessment, monitoring and imaging. *Nanomedicine* **2011**, *6*, 879–898. [CrossRef] [PubMed]
238. Mohamed El Mahdy, M.; Salah, T.; Sayed Aly, H.; Mohammed, F.; Shaalan, M. Evaluation of hepatotoxic and genotoxic potential of silver nanoparticles in albino rats. *Exp. Toxicol. Pathol.* **2015**, *67*, 21–29. [CrossRef] [PubMed]
239. Pinzaru, I.; Coricovac, D.; Dehelean, C.; Moaca, E.A.; Mioc, M.; Baderca, F.; Sizemore, I.; Brittle, S.; Marti, D.; Calina, C.D.; et al. Stable peg-coated silver nanoparticles—A comprehensive toxicological profile. *Food Chem. Toxicol.* **2018**, *111*, 546–556. [CrossRef] [PubMed]
240. Jacob, J.A.; Sivalingam, P.; Chen, B. Toxicological effects of silver nanoparticles. *Environ. Toxicol. Pharmacol.* **2015**, *40*, 729–732.
241. Franci, G.; Falanga, A.; Galdiero, S.; Palomba, L.; Rai, M.; Morelli, G.; Galdiero, M. Silver nanoparticles as potential antibacterial agents. *Molecules* **2015**, *20*, 8856–8874. [CrossRef] [PubMed]
242. Wu, F.; Harper, B.J.; Harper, S.L. Differential dissolution and toxicity of surface functionalized silver nanoparticles in small-scale microcosms: Impacts of community complexity. *Environ. Sci. Nano* **2017**, *4*, 359–372. [CrossRef]
243. Dubey, P.; Matai, I.; Kumar, S.U.; Sachdev, A.; Bhushan, B.; Gopinath, P. Perturbation of cellular mechanistic system by silver nanoparticle toxicity: Cytotoxic, genotoxic and epigenetic potentials. *Adv. Colloid Interface Sci.* **2015**, *221*, 4–21. [CrossRef] [PubMed]
244. Ribeiro, A.P.C.; Anbu, S.; Alegria, E.; Fernandes, A.R.; Baptista, P.V.; Mendes, R.; Matias, A.S.; Mendes, M.; Guedes da Silva, M.F.C.; Pombeiro, A.J.L. Evaluation of cell toxicity and DNA and protein binding of green synthesized silver nanoparticles. *Biomed. Pharmacother.* **2018**, *101*, 137–144. [CrossRef] [PubMed]
245. Lin, C.X.; Yang, S.Y.; Gu, J.L.; Meng, J.; Xu, H.Y.; Cao, J.M. The acute toxic effects of silver nanoparticles on myocardial transmembrane potential, ina and ik1 channels and heart rhythm in mice. *Nanotoxicology* **2017**, *11*, 827–837. [CrossRef] [PubMed]
246. McShan, D.; Ray, P.C.; Yu, H. Molecular toxicity mechanism of nanosilver. *J. Food Drug Anal.* **2014**, *22*, 116–127. [CrossRef] [PubMed]

247. Senthil, B.; Devasena, T.; Prakash, B.; Rajasekar, A. Non-cytotoxic effect of green synthesized silver nanoparticles and its antibacterial activity. *J. Photochem. Photobiol. B Biol.* **2017**, *177*, 1–7. [CrossRef] [PubMed]
248. Kora, A.J.; Sashidhar, R.B. Biogenic silver nanoparticles synthesized with rhamnogalacturonan gum: Antibacterial activity, cytotoxicity and its mode of action. *Arab. J. Chem.* **2018**, *11*, 313–323. [CrossRef]
249. Dos Santos, C.A.; Seckler, M.M.; Ingle, A.P.; Gupta, I.; Galdiero, S.; Galdiero, M.; Gade, A.; Rai, M. Silver nanoparticles: Therapeutical uses, toxicity, and safety issues. *J. Pharm. Sci.* **2014**, *103*, 1931–1944. [CrossRef] [PubMed]
250. Sudha, A.; Jeyakanthan, J.; Srinivasan, P. Green synthesis of silver nanoparticles using lippia nodiflora aerial extract and evaluation of their antioxidant, antibacterial and cytotoxic effects. *Resour.-Effic. Technol.* **2017**, *3*, 506–515. [CrossRef]
251. Furno, F.; Morley, K.S.; Wong, B.; Sharp, B.L.; Arnold, P.L.; Howdle, S.M.; Bayston, R.; Brown, P.D.; Winship, P.D.; Reid, H.J. Silver nanoparticles and polymeric medical devices: A new approach to prevention of infection? *J. Antimicrob. Chemother.* **2004**, *54*, 1019–1024. [CrossRef] [PubMed]
252. Gliga, A.R.; Skoglund, S.; Odnevall Wallinder, I.; Fadeel, B.; Karlsson, H.L. Size-dependent cytotoxicity of silver nanoparticles in human lung cells: The role of cellular uptake, agglomeration and Ag release. *Part. Fibre Toxicol.* **2014**, *11*, 11. [CrossRef] [PubMed]
253. Galbiati, V.; Cornaghi, L.; Gianazza, E.; Potenza, M.; Donetti, E.; Marinovich, M.; Corsini, E. In vitro assessment of silver nanoparticles immunotoxicity. *Food Chem. Toxicol.* **2018**, *112*, 363–375. [CrossRef] [PubMed]
254. Salama, A. Dicarboxylic cellulose decorated with silver nanoparticles as sustainable antibacterial nanocomposite material. *Environ. Nanotechnol. Monit. Manag.* **2017**, *8*, 228–232. [CrossRef]
255. Ivask, A.; ElBadawy, A.; Kaweeteerawat, C.; Boren, D.; Fischer, H.; Ji, Z.; Chang, C.H.; Liu, R.; Tolaymat, T.; Telesca, D.; et al. Toxicity mechanisms in *Escherichia coli* vary for silver nanoparticles and differ from ionic silver. *ACS Nano* **2014**, *8*, 374–386. [CrossRef] [PubMed]
256. Zhang, T.; Wang, L.; Chen, Q.; Chen, C. Cytotoxic potential of silver nanoparticles. *Yonsei Med. J.* **2014**, *55*, 283–291. [CrossRef] [PubMed]
257. Ivask, A.; Kurvet, I.; Kasemets, K.; Blinova, I.; Aruoja, V.; Suppi, S.; Vija, H.; Käkinen, A.; Titma, T.; Heinlaan, M.; et al. Size-dependent toxicity of silver nanoparticles to bacteria, yeast, algae, crustaceans and mammalian cells in vitro. *PLoS ONE* **2014**, *9*, e102108. [CrossRef] [PubMed]
258. Durán, N.; Silveira, C.P.; Durán, M.; Martinez, D.S.T. Silver nanoparticle protein corona and toxicity: A mini-review. *J. Nanobiotechnol.* **2015**, *13*, 55. [CrossRef] [PubMed]
259. Nguyen, K.C.; Seligy, V.L.; Massarsky, A.; Moon, T.W.; Rippstein, P.; Tan, J.; Tayabali, A.F. Comparison of toxicity of uncoated and coated silver nanoparticles. *J. Phys. Conf. Ser.* **2013**, *429*, 012025. [CrossRef]

nanomaterials

MDPI

Article

Bifunctionalized Silver Nanoparticles as Hg^{2+} Plasmonic Sensor in Water: Synthesis, Characterizations, and Ecosafety

Paolo Prosposito [1,2], Luca Burratti [1], Arianna Bellingeri [3], Giuseppe Protano [3], Claudia Faleri [4], Ilaria Corsi [3], Chiara Battocchio [5], Giovanna Iucci [5], Luca Tortora [5,6], Valeria Secchi [5], Stefano Franchi [7] and Iole Venditti [5,*]

[1] Department of Industrial Engineering and INSTM, University of Rome Tor Vergata, via del Politecnico 1, 00133 Rome, Italy; paolo.prosposito@uniroma2.it (P.P.); luca.burratti@uniroma2.it (L.B.)
[2] Center for Regenerative Medicine, University of Rome Tor Vergata, Via Montpellier 1, 00133 Rome, Italy
[3] Department of Physical, Earth and Environmental Sciences, University of Siena, Via Mattioli 4, 53100 Siena, Italy; arianna.bellingeri@student.unisi.it (A.B.); giuseppe.protano@unisi.it (G.P.); ilaria.corsi@unisi.it (I.C.)
[4] Department of Life Sciences, via Mattioli 4, 53100 Siena, Italy; faleric@unisi.it
[5] Department of Sciences, Roma Tre University of Rome, Via della Vasca Navale 79, 00146 Rome, Italy; chiara.battocchio@uniroma3.it (C.B.); giovanna.iucci@uniroma3.it (G.I.); luca.tortora@uniroma3.it (L.T.); valeria.secchi@uniroma3.it (V.S.)
[6] Surface Analysis Laboratory INFN Roma Tre, via della Vasca Navale 84, 00146 Rome, Italy
[7] Elettra-Sincrotrone Trieste S.c.p.A., Strada Statale 14, km 163.5, 34149 Basovizza Trieste, Italy; stefano.franchi@elettra.eu
[*] Correspondence: iole.venditti@uniroma3.it; Tel.: +39-06-5733-3388

Received: 20 August 2019; Accepted: 16 September 2019; Published: 20 September 2019

Abstract: In this work, hydrophilic silver nanoparticles (AgNPs), bifunctionalized with citrate (Cit) and L-cysteine (L-cys), were synthesized. The typical local surface plasmon resonance (LSPR) at λ_{max} = 400 nm together with Dynamic Light Scattering (DLS) measurements ($<2R_H> = 8 \pm 1$ nm) and TEM studies (Ø = 5 ± 2 nm) confirmed the system nanodimension and the stability in water. Molecular and electronic structures of AgNPs were investigated by FTIR, SR-XPS, and NEXAFS techniques. We tested the system as plasmonic sensor in water with 16 different metal ions, finding sensitivity to Hg^{2+} in the range 1–10 ppm. After this first screening, the molecular and electronic structure of the AgNPs-Hg^{2+} conjugated system was deeply investigated by SR-XPS. Moreover, in view of AgNPs application as sensors in real water systems, environmental safety assessment (ecosafety) was performed by using standardized ecotoxicity bioassay as algal growth inhibition tests (OECD 201, ISO 10253:2006), coupled with determination of Ag^+ release from the nanoparticles in fresh and marine aqueous exposure media, by means of ICP-MS. These latest studies confirmed low toxicity and low Ag^+ release. Therefore, these ecosafe AgNPs demonstrate a great potential in selective detection of environmental Hg^{2+}, which may attract a great interest for several biological research fields.

Keywords: silver nanoparticles; Hg^{2+} sensors; heavy metal sensing; plasmonic sensors; optical sensors; ecosafety

1. Introduction

Nanosized inorganic particles, in simple or composite formulation, have unique physical and chemical properties and represent an increasingly important material in the development of novel nanodevices that can be used in numerous fields such as catalysis, energy, optoelectronics, biomedicine, and sensors [1–11]. Several recent achievements show the opportunity of producing new types of nanostructured materials with planned surface and desired physico-chemical properties [12–16]. Among other materials, silver nanoparticles (AgNPs) are deeply studied for their optical and antibacterial properties, easy functionalizations and cheap preparations [17,18].

Nowadays, for the AgNPs use in sensing field, the main goal is the preparation of uniform nanosized particles with precise requirements in terms of size, shape, surface functionalities, and structural features [19–22]. In fact, many studies involved AgNPs as optical sensor, using the localized surface plasmon resonance (LSPR) that is specific feature of the colloidal silver nanoparticle solutions: the energy of the absorption maximum and the shape of the peak are strongly related with the size, shape and interparticle distance, but also with the surrounding environment, which can influence the degradation or aggregation of the particles [23–25]. These optical properties of NPs have allowed researchers to mature new diagnostic methods that are useful for optical and colorimetric measurements [26–28].

In particular, AgNPs have drawn great attention for mercury sensors development, especially due to their high sensitivity to Hg^{2+} micro-level concentration changes [29,30]. This behavior has been related, both with redox chemistry of Hg^{2+} and silver surface (Ag°), and both with the soft-soft chemistry between metals and sulfur-containing capping agents used for AgNPs [31]. In both cases, the mercury presence produce important changes in the absorbance intensity and peak position of AgNPs. Surface functionalization has the main role both in stabilization and in reactivity of AgNPs. Therefore, the ligand surface chemistry is used to tune their sensing properties. For example, many ligands, such as *N*-cholyl-Lcysteine [32], citrate [33–35], 6-thioguanine [36], cytosine triphosphate [37], have been used for the surface modifications of AgNPs and AuNPs, and the different functionalizations change the inter and intra-ligand interactions. Obviously, all this produces different properties and capacities of the nanoparticles, that can aggregate or disaggregate and interact with elements and ions present in their surroundings, and it affects sensing features towards metals ions, such as Hg^{2+}, modifying the selectivity and sensitivity. So far there are several reports available for the colorimetric detection of Hg^{2+} ions using green synthesized unmodified AgNPs in aqueous medium [38]. Frequently used, citrus extracts from lemons and sweet orange fruits have been investigated for green synthesis of AuNPs and AgNPs used as Hg^{2+} colorimetric sensors [39]. There are many bio extracts reported from different plants, fruits, leafs, etc. which have been used for photoinduced green synthesis of AgNPs where a bio extract acts as reducing as well as stabilizing agent and applied in Hg^{2+} detection [38–40].

Nevertheless, the use of NPs for heavy metals removal known as nanoremediation is a matter of concern due to the potential risks associated with the scarcity of information regarding their behavior, fate and impact on the environment and human health [41]. In order to overcome such limitations, an ecosafe approach is developed with the aim to test the ecotoxicity of NPs to selected biota having an important ecological role in aquatic ecosystems, as for instance primary producers such as freshwater and marine microalgae [42,43].

AgNPs are recognized to exert toxic effects to biota, both from fresh- and marine waters, from bacteria [44] and microalgae [45–48] to crustacean [49], mussels [50], fishes [51] and mammalian cells [52,53]. Among these studies, those reporting Ag^+ release in exposure media [44–48,52,54], report that AgNPs toxicity is often closely linked to the dissolution of the particles and consequent release of Ag ions during exposure. Therefore, based on such eVidence, the ecosafety assessment of AgNPs is mandatory in order to avoid any potential toxicological risks associated with their application in environmental remediation.

In this work AgNPs functionalized with hydrophilic capping agents, i.e., citric acid (Cit) and L-Cysteine (L-cys) were synthesized and characterized by means of UV-Vis, Fourier Transform Infrared (FTIR), Synchrotron radiation-X-ray photoelectron (SR-XPS), Near Edge X-ray Absorption Fine Structure (NEXAFS) spectroscopies. Their nanodimensions were confirmed by Dynamic Light Scattering (DLS) and Transmission Electron Microscope (TEM) analysis. AgNPs were tested as plasmonic sensor for heavy metal detection in water, showing selectivity and sensitivity for Hg^{2+} ions. The AgNPs-Hg^{2+}

system was deeply studied by means of UV-Vis, and, SR-XPS spectroscopies, DLS, Inductively Coupled Plasma-Mass Spectrometry (ICP-MS) and TEM studies. Moreover, in view of AgNPs application in real water systems, their ecosafety was investigated.

2. Experimental

2.1. Materials and Methods

Sodium citrate ($Na_3C_6H_5O_7$, Cit), L-Cysteine ($C_3H_7NO_2S$, L-cys), silver nitrate ($AgNO_3$) and sodium borohydride ($NaBH_4$) have been used as received (reagent grade, Sigma-Aldrich, St. Louis, MO, USA). Metal ions contamination was accomplished by using the following salts:

$NaAsO_2$, $NaHAsO_4{\cdot}7H_2O$, $Ca(ClO_4)_2$, $Cd(NO_3)_2$, $CoCl_2{\cdot}6H_2O$, $CrCl_3{\cdot}6H_2O$, $Cu(NO_3)_2$, $FeCl_3{\cdot}6H_2O$, $Hg(NO_3)_2{\cdot}H_2O$, $KClO_4$, $Mg(ClO_4)_2$, $NaClO_4$, $NdCl_3{\cdot}6H_2O$, $NiCl_2{\cdot}6H_2O$, $Pb(NO_3)_2$, $Zn(NO_3)_2{\cdot}6H_2O$. For all the solutions, we used deionized water (electrical conductivity less than 1 μΩ/cm at room temperature) obtained from Millipore Milli-Q water purification system. All the reagents were purchased (from Sigma-Aldrich, St. Louis, MO, USA) and were used without further purification.

UV-Vis spectra were run in H_2O solution by using quartz cells with a Shimadzu 2401 PC UV-vis spectrophotometer and by using single-use UV-PMMA cuvettes with Perkin-Elmer Lambda 19 UV-Vis-NIR for sensing test characterization. ATR-FTIR spectra have been recorded for films deposited by casting from water suspension with an FTIR spectrometer (Nicolet iS50, Thermo Fisher Scientific, Madison, WI, USA) equipped with a mid- and far-IR capable diamond ATR accessory. FT-IR spectra were recorded in the range between 350 and 4000 cm^{-1} with a resolution of 4 cm^{-1}, a zero-filling factor of 2 and the co-addition of 64 scans. Data were acquired using OMNIC software (version 9.8.372, Thermo Scientific), subtracting the air background spectrum obtained prior to each sample spectrum acquisition. AgNPs were investigated by TEM (Philips Morgagni 268 D electronics, at 80 KV and equipped with a MegaView II CCD camera, Philips Electronics, Eindhoven, The Netherlands) at 10 mg/L in MilliQ water. The obtained images were analysed with *ImageJ* for particles size measurement. A total of 180 particles were measured in two portions of two different images and the average size was calculated. Size distribution of AgNPs (50 mg/L) have been investigated by means of DLS (Zetasizer Nano Series, Malvern instruments, Enigma Business Park, Grovewood Rd, UK), combined with the Zetasizer Nano Series software (version 7.02, Particular Sciences) at T = 25.0 ± 0.2 °C in milliQ water, as well as media used for toxicity assessment (freshwater, TG 201, and marine water, F/2). Correlation data have been acquired and fitted in analogy to our previous work [55,56]. Ag^+ release from the AgNPs has been assessed in both TG 201 and F/2 aqueous media, at 0 h and after 72 h. Six solutions have been prepared: TG 201, F/2, TG 201 + 500 μg AgNPs/L, F/2 + 500 μg AgNPs/L, TG 201 + 7 μg $AgNO_3$/L, F/2 + 7 μg $AgNO_3$/L. Solutions were kept in the same conditions as toxicity tests (22 ± 2 °C and 16/8 light-dark photoperiod) and mixed by shaking once a day. An aliquot of each solution was taken at 0 h and after 72 h and centrifuged (5000 g, 40 min, 22 °C) by using a centrifugal filter device with a 3 kDa cut-off (Amicon Ultra-15 mL, Millipore, Sigma-Aldrich, St. Louis, MO, USA). The resulting filtrate was acidified with HNO_3 (10%) and analysed by ICP-MS (Perkin Elmer NexION 350 spectrometer) for determining Ag concentration.

SR-XPS measurements were performed at the Elettra synchrotron radiation source (Trieste, Italy), using the Materials Science Beamline (MSB), that is positioned at the left end of the bending magnet 6.1. MSB is equipped with a plane grating monochromator providing SR light in the 21–1000 eV energy range. The UHV endstation, whose base pressure is of 2×10^{-10} mbar, is equipped with a SPECS PHOIBOS 150 hemispherical electron analyser, low-energy electron diffraction optics, a dual-anode Mg/Al X-ray source, an ion gun, a sample manipulator with a K-type thermocouple attached to the rear side of the sample. For the here presented experiments we detected photoelectrons emitted by C1s, O1s, S2p, Ag3d, N1s and Hg4f core levels, using a normal emission geometry. In order to maximize signals intensity, we selected a photon energy value of 630 eV (impinging at 60°) for all elements except S; in order to maximize the intensity of S2p signals, that was expected to be very low due to element dilution, the S2p core level was measured with photon energy = 350 eV. Charging correction of binding energies (BEs) was done using as a reference the aliphatic C1s (BE 285.0 eV) [57]. To fit core level spectra, we subtracted a Shirley background and then used Gaussian peak functions as signals components [58,59].

NEXAFS experiments were carried out at the BEAR (Bending magnet for Emission Absorption and Reflectivity) beamline, installed at the left exit of the 8.1 bending magnet and located the ELETTRA third generation storage ring. The beamline optics can deliver photons having energy comprised between 5 eV and 1600 eV with selectable degree of ellipticity. The carbon K-edge spectra were recorded at normal (90°) and grazing (20°) incidence relative to the sample surface of the linearly polarized photon beam; however, no angular effects were detected. Calibration of the photon energy and resolution was carried out at the K absorption edges of Ar, N_2 and Ne. In order to normalize the spectra, a straight line fitting the part of the spectrum below the edge was subtracted and the value recorded at 330.00 eV was assessed to 1.

2.2. AgNPs Synthesis

The AgNPs stabilized with citrate (Cit) and L-Cysteine (L-cys) were prepared and characterized in analogy to literature reports [60,61]. 1.47 g of sodium citrate were dissolved in 50 mL of distilled water (0.01 M), 0.006 g of L-cys in 25 mL of distilled water (0.002 M) and 0.21 g of $AgNO_3$ in 25 mL of distilled water (0.05 M). Then, 25 mL of L-cys solution, 10 mL of Cit solution and 2.5 mL of $AgNO_3$ solution were added sequentially in a 100 mL flask, provided with a magnetic stir. The mixture was degassed with Argon for 10 min, then 4 mL of $NaBH_4$ solution (0.016 g in 4 mL distilled water) were added. The mixture was allowed to react at room temperature for 2 h and at the end the brown solution was recollected and purified by centrifugation (13,000 rpm, 10 min, 2 times with deionized water). AgNPs main characterizations: UV-Vis (λ_{max} [nm], H_2O) = 400 nm; $< 2R_H >$ ([nm], H_2O) = 8 ± 1; TEM Ø = 5 ± 2 nm.

2.3. AgNPs Sensing Tests

A fixed volume of AgNPs in water (typical concentration 1.6 mg/mL) was mixed with a fixed volume of water solution containing the heavy metal ions at specific concentration. After five minutes of interaction of the nanoparticles with the metal ions, the optical absorption spectra were collected in order to verify possible changes (shape, energy and intensity) of the typical localized surface plasmon resonance (LSPR). The response to several metal ions at different concentrations from 25 ppm down to 1 ppm was tested by UV-Vis spectroscopy [27].

2.4. AgNPs Ecosafety Tests

The intrinsic toxicity of AgNPs was investigated through standardized 72 h algal growth inhibition tests [62,63] using two different algal species, belonging to the freshwater and to the marine water environment, *Raphidocelis subcapitata* and *Phaeodactylum tricornutum*. Algae were cultured, respectively, in TG 201 medium and F/2 medium, in axenic and exponential growth conditions, at 18 ± 1 °C and 16/8 h light-dark cycle photoperiod. In order to reduce the introduction of chelating agents which have been shown to interfere with heavy metals toxicity [64,65], algal toxicity tests were run with the same medium used for algal culturing modified only in the concentration of ethylenediaminetetraacetic acid (EDTA) (0.05 mg/L and 0.8 mg/L respectively as previously proved to ensure acceptable algal growth). PS single-use sterile multiwell were used for toxicity tests, with 2 mL volume capacity for each well. Algae were acclimated 72 h before running the test under the following conditions: 22 ± 2 °C and 16/8 light-dark photoperiod. Initial algal concentration was 1×10^4 cells/mL and tests were manually aerated eVery 24 h using a pipette with sterile tips. Exposure concentrations were as follows: 10, 25, 50, 100, 200, 500 μg AgNPs/L. $AgNO_3$ was used as positive control at the following concentrations of 3.5, 7,

14, 21, 35 μg/L. In order to exclude any possible role of coating components the toxicity of the AgNPs, citrate and L-Cysteine were tested separately at the following concentrations of 0.5, 1, 2, 5, 10 mg/L.

3. Results and Discussion

3.1. AgNPs Synthesis and Characterizations

The synthesis of AgNPs by wet chemical reduction is a useful method to obtain spherical nanoparticles with tuned sizes and opportune capping agent [60]. In this work, AgNPs have been prepared by the reduction in aqueous solution of silver nitrate with sodium borohydride as strong reducing agent. Two different capping agents have been chosen: Cit, crucial to induce a high hydrophilic behavior, and L-cys, to induce selective interaction with the environment.

In fact, it is well known that the amino group can easily interact with the chemical environment and in particular with Hg^{2+} ions [10]. In order to avoid the excessive presence of L-cys, leading to nanoparticles aggregation, the molar ratio Au/Cit/L-cys = 1/4/2 was chosen. The obtained AgNPs scheme is shown in Figure 1a. After careful purification, AgNPs have been characterized by means of UV-Vis, FTIR and XPS spectroscopies, by DLS and TEM studies. UV-Vis spectra have been carried out together with DLS measurements in water, as reported in Figure 1b,c. The UV-Vis spectrum showed the typical LSPR band, at λ_{max} = 400 nm, confirming the nanodimensions. DLS measurements in water showed hydrodynamic diameter $<2R_H>$ = 8 ± 1 nm, as expected. Moreover, FTIR investigations showed the presence of Cit and L-cys on the particles, as reported in Figure 1d.

When the ATR-FTIR spectrum of AgNPs bifunctionalized with Cit and L-cys was compared with reference spectra of these capping agents Figure 1d, it is immediately eVident the absence of the band at ~1582 cm^{-1} ($\nu_{as}(COO^-)$) in AgNPs spectrum [66]. This could be interpreted as eVidence that most carboxylate groups of the citrate and cysteine are attached to the surface of the AgNPs. In fact, similar results was observed by Frost et al. [67], who studied the ATR-FTIR spectra of citrate-capped AgNPs and Au-NPs; only for AgNPs, they observed the disappearance of the $\nu_{as}(COO^-)$ peak and interpreted this result in terms of adsorption geometry of the citrate molecule, with all the three carboxylate groups interacting with the AgNPs surface: the $\nu_{as}(COO^-)$ lying parallel to the AgNPs surfaces results in an infrared-inactive transition. Moreover, in the AgNPs spectrum is present a broad peak at ~1547 cm^{-1} that could be due to the overlapping between the N-H bending of cysteine and the residual carboxylate groups not attached to the AgNPs surface. The functionalization with Cit and L-cys can be also supported by the presence in the AgNPs spectrum of a broad peak centered at 1379 cm^{-1} indicating the COO^- symmetric stretching vibrations; the peak position is slightly shifted to lower wavenumber compared to the free carboxylate anion, as already eVidenced by Frost et al. [68]. The main contributions to the symmetric and asymmetric stretching vibration of the carboxylate group are obviously mainly due to the citrate molecule, that is present in higher concentration on the AgNPs; however, it seems reasonable to hypothesize a similar reactivity for the different carboxylate groups. The broad peak centered at 3300 cm^{-1} is attributed to the stretching vibrations of the hydroxyl group (ν(O–H)) (data not shown) probably coming from water adsorbed onto nanoparticles surface. We were not able to detect any S-H band around 2550 cm^{-1}, that would indicate the presence of free L-cys on the AgNPs surface [68,69]. The intensity of this peak in the spectrum of L-cys is actually rather low (see Figure S2 in Supplementary Material), however, XPS results confirm the absence of S-H groups.

To support FTIR data, AgNPs were also investigated by NEXAFS spectroscopy. C k edge spectra of AgNPs were recorded at normal, magic and grazing incidence; however, no angle-dependent effect was detected; therefore, the C k-edge spectrum of AgNPs is shown in Figure 2. The main feature in the spectrum is the 1s→π^* consisting of two peaks located at 287.7 and 288.8 eV, with a shoulder at 285.5 eV. According to literature, the 1s→π^* transition related to the C=O bond in the carboxylate of L-cys is expected at 288.6 eV [70]; similar values are expected for carboxyl groups [71]. Moreover, L-cys is expected to show a 1s→σ^* peak related to the C-S bond at about 287.3 eV. Therefore, we can assign this complex band to overlap between π^*C=O and σ^*C–S transitions. The broad bands located at about

294 and 300 eV are related to 1s→σ* transitions of C–H and C=O bonds respectively [72]. The overall spectrum yields prove of successful capping of the AgNPs surface by both Cit and L-cys.

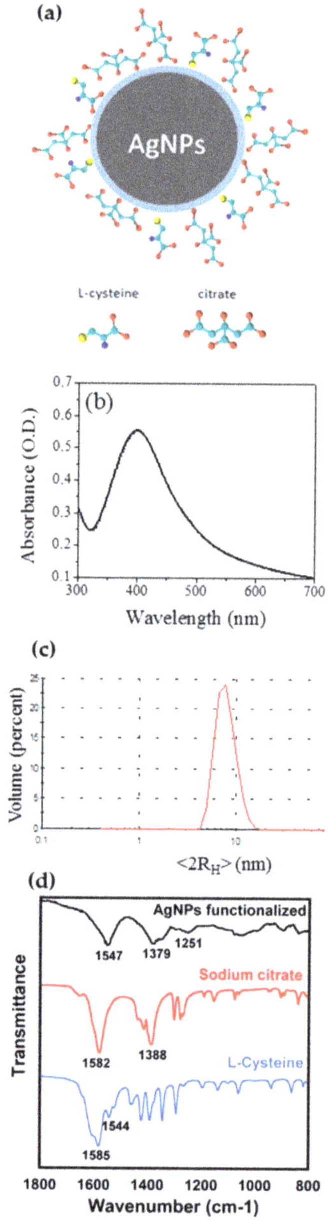

Figure 1. (**a**) Scheme of bifunctionalized hydrophilic silver nanoparticles (AgNPs); (**b**) UV-Vis spectrum in water of AgNPs, with local surface plasmon resonance (LSPR) band centred at 400 nm; (**c**) DLS measurements in water: <2RH> = 8 ± 1 nm; (**d**) ATR-FTIR spectra of bifunctionalized AgNPs (Top) and the capping agents, Cit (Center), and L-cys (Bottom).

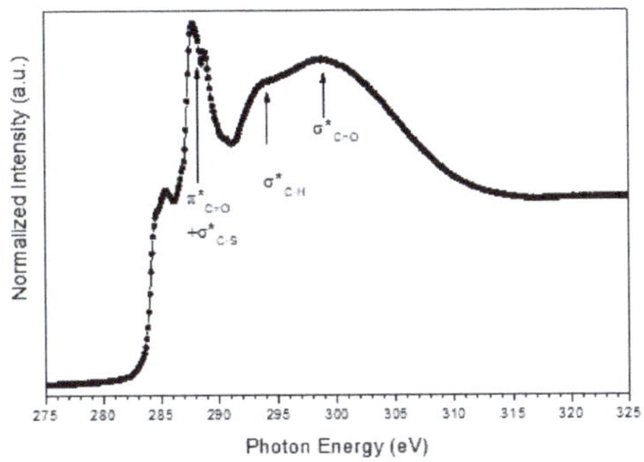

Figure 2. C k-edge NEXAFS spectrum of AgNPS recorded at 20° incidence angle.

Moreover, the TEM studies carried out on diluted samples show AgNPs with narrow size distribution (TEM Ø 5 ± 2 nm) but with a certain tendency to aggregation, which could be due only to drying (see Figure S1). DLS measurements show a monodisperse population of NPs in solution These data are in agreement with DLS data. It can be noticed that the dimensions obtained from DLS are greater than those obtained from TEM images, since DLS estimated the hydrodynamic radius $< 2R_H >$ of the particles in the aqueous environment and it is the Z-average value, which is the mean diameter weighted over the scattered light intensity. Therefore, the DLS data are not directly comparable with dimensions obtained from TEM images, as reported in the literature [7,56]. Anyway, the data confirmed in both cases the nanodimension of the AgNPs.

Measurements of Ag concentration in fresh and marine aqueous media showed an almost absent Ag^+ ions release from AgNPs (Table 1) as opposed to what observed in the literature for other types of AgNPs [45–49,53,55].

Table 1. Ag^+ concentrations (expressed as μg/L) in freshwater and marine waters with algal medium solution (CTRL), as well as algal medium solutions with AgNPs (500 μg/L) and with $AgNO_3$ (7 μg/L).

	TG 201 (Freshwater)		**F/2 (Marine Water)**	
	0 h	**72 h**	**0 h**	**72 h**
CTRL	0.36 ± 0.06	0.26 ± 0.08	0.27 ± 0.01	0.26 ± 0.02
AgNPs	0.19 ± 0.02	0.4 ± 0.06	0.15 ± 0.03	0.37 ± 0.03
$AgNO_3$	4.42 ± 0.09	4.79 ± 0.18	4.37 ± 0.08	5.32 ± 0.22

In fact, Ag^+ concentrations in both TG 201 medium (freshwater) and F/2 medium (marine water) with 500 μg AgNPs/L were low (up to 0.4 μg/L) and comparable with those measured in controls (up to 0.36 μg/L; Table 1). However, it is to be noted that in the presence of AgNPs, Ag^+ levels slightly increased after 72 h, from 0.19 to 0.4 μg/L in freshwater, and 0.15 to 0.37 μg/L in marine water. Since AgNPs dissolution was demonstrated to be independent from particles aggregation state [73], the reason for the lack of ion release is probably to be found in the Cit/L-cys coating.

3.2. AgNPs Sensing Tests

Sensing tests were made by checking the optical absorption spectra of the contaminated systems with respect to the reference AgNPs water solution (see also Figure S3). In Figure 3a the absorption spectra of AgNPs in water solutions without (reference solution) and with different concentrations of Hg^{2+} ions in the range 1 to 10 ppm, are reported. An eVident red shift of the band peak energy together with an increase of the intensity and a broadening of the band with the increasing concentration of ions can be appreciated. The same type of measurement has been performed for all the 16 ions listed in the experimental part for concentration from 10 ppm down to the lowest concentration of 1 ppm.

Figure 3b shows the red shift of the absorption band of all the metal ions tested at the concentration of 2.5 ppm. A significant change has been detected only for Hg^{2+} ions while for all the other ions the absorption bands remain almost constant within the error bars.

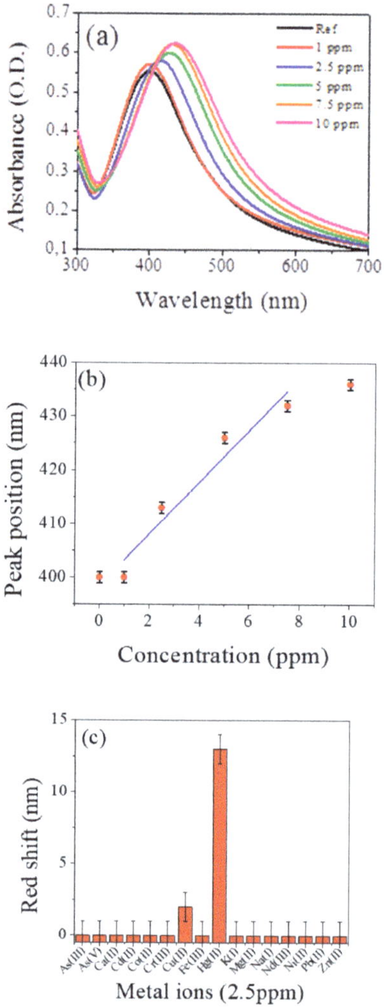

Figure 3. (**a**) Absorption spectra of AgNPs water solutions at room temperature and pH = 6.5 without and with different concentrations of Hg^{2+} listed in the figure; (**b**) calibration curve as a function of Hg^{2+} concentration; (**c**) redshift of the absorption band maximum in presence of all the metal ions tested at concentration of 2.5 ppm.

The limit of detection (LOD) of the nanosensor for the analysis of Hg^{2+} was determined using calibration curves. The LOD can be estimated as three times the standard deviation of the blank signal, obtaining a value equal to 0.6 ppm. The obtained LOD was found to be comparable with other AgNPs systems and the experimental results in the determination of Hg^{2+} obtained by some other methods are listed in Table 2 for comparison [8,10,40,74–77].

Table 2. Comparison between dimension and limit of detection (LOD) of some metal NPs used as a colorimetric sensor for the detection of Hg^{2+}.

Metal NPs Functionalization	NPs Diameter (nm)	Hg^{2+} Detection Limit (LOD) (M)	References
Citrate/Lcysteine-AgNPs	5 ± 2	3.0×10^{-6} (0.6 ppm)	This work
Tween 20-AuNPs	-	1.0×10^{-7}	[8]
L-cys-AuNPs	-	1.0×10^{-7}	[10]
Acanthe phylum bracteatum AgNPs	29–68	2.2×10^{-6}	[40,74]
L-cysteine AgNPs	10	1.0×10^{-8}	[75]
Glutamine/Histidine AgNPs	5.5 ± 1.0	9.0×10^{-7}	[76]
Glutathione AgNPs	3.9 ± 0.6	$>1.0 \times 10^{-7}$	[77]

3.3. SR-XPS Characterization of AgNPs and AgNPs-Hg^{2+}

Synchrotron radiation-induced X-ray photoelectron spectroscopy (SR-XPS) measurements allowed to probe the interaction between AgNPs and Hg^{2+} ions. Spectra were collected at C1s, N1s, O1s, Ag3d, S2p and Hg4f core levels (all BE (eV), FWHM (eV), relative intensity values and proposed signals assignments are reported in Supplementary Materials file, Table S1). All the individuated spectral components confirm AgNPs stability, and Cit and L-cys capping efficiency. C1s spectrum has three components at 285.00, 286.46 and 288.45 eV BE, respectively associated with aliphatic carbons (mainly impurities, that are always observed in samples prepared in air by liquid solutions); O1s spectra, reported in Figure 4c, also show three different kinds of oxygen atoms, respectively belonging to carbonyl (C=O) functional groups (532.00 eV BE), hydroxyl moieties (-OH, 533.00 eV BE) and physisorbed water (small contribution at about 534.5 eV BE). The atomic ratio between C=O and –OH is C=O/–OH = 1/1.4, very close to the C=O/–OH = 1/1.3 that is theoretically expected for a Cit/L-cys stoichiometry = 2/1, as in the synthetic procedure reported in Materials and Methods. N1s spectrum was also collected, showing a single component centred at 400.24 eV, as expected for amine-like nitrogens. Indeed, C1s, O1s, and N1s data analysis confirm the molecular stability of the AgNPs, also after interaction with Hg^{2+} ions.

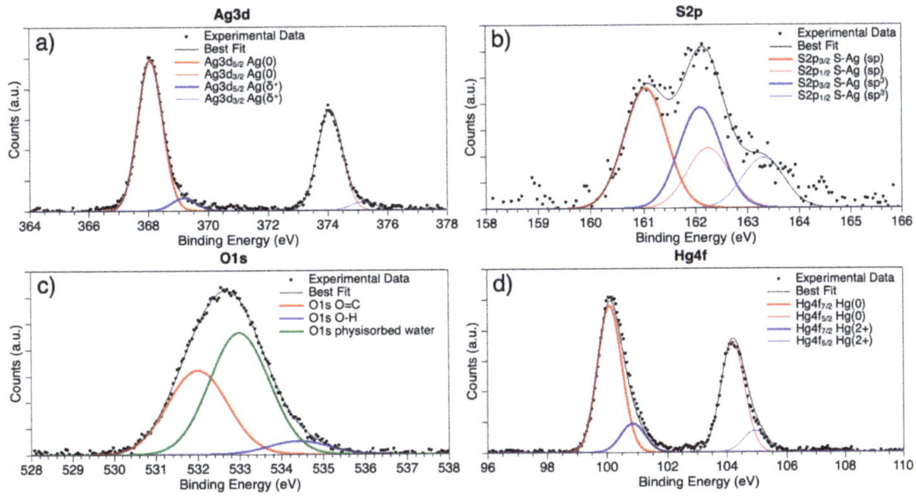

Figure 4. SR-XPS spectra collected on AgNPs-Hg^{2+} aggregates at (**a**) Ag3d; (**b**) S2p; (**c**) O1s and (**d**) Hg4f core levels.

Generally speaking, the most indicative signals for the surface-structure analysis of metal nanoparticles capped with thiols are M (Au4f, Ag3d) and S2p core levels; for this purpose, Ag3d and S2p spectra are reported in Figure 4 and will be here discussed in detail.

Ag3d spectra Figure 4a are asymmetric at high BE, a common feature in capped nanoparticles [20,27], indicating that at least two different kinds of silver atoms compose the nanoparticle: the spin-orbit pair at lower BE values (Ag3d5/2 = 368.08 eV) is associated with metallic silver at the NPs core; the signal at higher BE, of very small intensity (about 9% of the whole signal) is due to positively charged silver atoms at the NP surface, interacting with the capping molecule [20,27]. S2p spectra are also composite, showing two spin orbit pairs of very similar intensity (atomic percents are 54.4% lower BE–45.6% higher BE, as reported in Table S1 in the Supplementary Material); interestingly, the two signals are both indicative for sulphur atoms covalently bonded to silver, but with two different hybridizations: the spin-orbit pair at lower BE (S2p3/2 = 161.05 eV) is indicative for S-Ag bonds with sp hybridized sulphur; the signal at higher BE (S2p3/2 = 162.09 eV) suggests S-Ag bonds with sp3 S atoms [78]. It is noteworthy that no physisorbed thiol moieties appear (R-SH S2p3/2 signals are expected around 163–164 eV BE); this finding is in excellent agreement with the hypothesis made by A Majzik et al. [61] based on ^{1}H NMR studies and suggesting that in metal nanoparticles stabilized by mixed Cit and L-cys capping agents the L-cys molecules preferentially tend to directly bond the metal surface, inducing the most part of Cit molecules to form a shell around the first L-cys layer, in a "layer-by-layer"-like arrangement. It is noteworthy that XPS data do not allow excluding that some Cit molecules could intercalate between L-cys and directly interact with Ag atoms at the NP surface, as eVidenced by IR spectroscopy results. The last signal reported in Figure 4d is the Hg4f spectrum. As eVidenced in the figure, two spin-orbit pairs can be individuated; the first signal ($Hf4f_{7/2}$ BE = 100.10 eV) can be associated with metallic Hg atoms [79]; the components at higher BE ($Hg4f_{7/2}$ = 100.85 eV) are consistent with literature data reported for Hg^{2+} ions in oxides or coordination compounds [72].

The occurrence of a signal indicative for metallic Hg is in excellent agreement with the findings reported in [68], where for AgNPs stabilized by Cit molecules and interacting with Hg(II) ions, a direct interaction between Hg and Ag atoms at the nanoparticle surface was envisaged. To better understand the chemistry and geometry of Hg(II)–AgNPs interaction, XAS experiments at the Hg L_{III}-edge are in programme. Actually, X-ray absorption experiments will allow to directly probe the local coordination chemistry of the metal ion, providing information complementary to the SR-XPS data and allowing for a precise description of the Hg coordination site.

3.4. AgNPs Ecosafety

Algal toxicity tests with *R. subcapitata* and *P. tricornutum* exposed to AgNPs showed no inhibition of growth rate at all tested concentrations, as reported in Figure 5. Furthermore, algae showed a slight increase in growth rate compared to the control (negative growth rate inhibition), at almost all tested concentrations. Such increase, however, was not observed by positive control tests carried out with Cit and L-cys, which showed no effect on algal growth.

Reference toxicant ($AgNO_3$) showed a 60% inhibition of growth rate compared to the control of *R. subcapitata* (freshwater), already at the lowest concentration tested (3.5 µg $AgNO_3$/L), confirming the toxicity of Ag for microalgae. On the opposite, no effect on growth was observed for *P. tricornutum* (marine water) exposed in the same conditions, confirming the of drop in free Ag^+ ions in solution due to complexation with Cl^- ions in seawater as measured by Gunsolus et al. [73].

The absence toxicity for our AgNPs confirms chemical analysis results showing insufficient release of Ag^+ ions to exert a toxic effect on both microalgae (Table 1). Such findings further validate the hypothesis by which AgNP toxicity is closely bound to the release of Ag^+ ions [44,46,47,54]. In fact, the dissolution of Ag^+ from AgNPs, either pristine or coated (chitosan, lactate, polyvinylpyrrolidone, polyethelene glycol, gelatin, sodium dodecylbenzenesulfonate, Cit, dexpanthenol, carbonate), has been reported by many studies [44–48,52,54,80,81] and recognized to be linked to the observed toxicity to microalgae [44,46,47,54]. The covalently-bound L-cys coating of AgNPs might prevent the dissolution of Ag^+ ions, since previous studies revealed that Cit coating was not able to avoid such release in aqueous media [44,46,47,73].

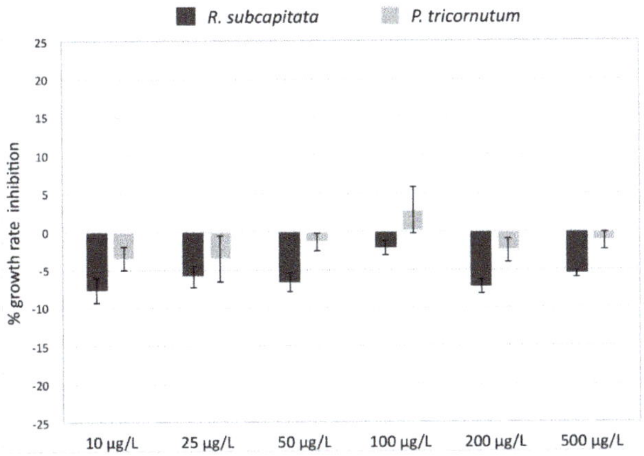

Figure 5. Percentage of growth rate inhibition compared to control of *R. subcapitata* and *P. tricornutum* exposed to AgNPs (10, 25, 50, 100, 200, 500 µg/L) for 72 h. Data are shown as mean ± standard deviation.

Molecules with a high reduced sulphur content are known to bind metal ions; L-cys, thanks to the presence of a thiol group, is able to act as a strong Ag^+ ligand [82]. Thiol groups could act both either binding Ag^+ ions, blocking their release in the media, or by excluding oxidizing agents to come in contact with the particles surface, preventing their dissolution and consequent release of Ag^+ ions in the medium [83]. Gunsolus et al., [73] reported a significant reduction of Ag^+ ions release and of bactericidal activity of citrate-AgNPs upon incubation with natural organic matter rich in thiol groups. Some studies observed a reduction in the toxic effect of Ag-releasing AgNPs after addition of L-cys in solution, probably as a consequence of L-cys complexation of free Ag^+ ions [45–47].

However, information on L-cys effects on AgNPs stability and dissolution are conflicting. The presence of free L-cys in solution might also enhance Ag^+ ions release from AgNPs [84]. In our study, however, the absence of toxicity to microalgae and low Ag levels in exposure media confirmed that Cit/L-cys coating prevents the release Ag^+ ions from the NP. Similar results were obtained by Mu et al. [85] for graphene oxide (GO) nanosheets with covalently-bound L-cys, compared to pristine GO nanosheets. In terms of toxicity, for particles smaller than 20 nm an additional dissolution-independent effect seems to play a role [44,53]. However, despite being very small, AgNPs further confirm their integrity in terms of low Ag^+ release. Since the main concern regarding the environmental application of AgNP is due to potential toxic effect to non-target aquatic species [44], our findings clearly showed the ecosafety of AgNPs and promote their use in the aquatic environment.

4. Conclusions

In this work AgNPs functionalized with hydrophilic capping agents, i.e., Cit and L-cys, were synthesized and characterized by means of UV-Vis, FT-IR, SR-XPS and NEXAFS spectroscopies, confirming the surface functionalization. Their nanodimensions were studied by DLS and TEM analysis, showing diameter less than 10 nm. AgNPs showed high selectivity and sensitivity for Hg^{2+} in water (concentration range:1–10 ppm) respect to 16 different metal ions investigated. The AgNPs-Hg^{2+} system was deeply investigated by means of DLS, ICP-MS, TEM and SR-XPS. Measurements of Ag concentration in fresh and marine aqueous media showed low Ag^+ ions release, probably due to the good Cit/L-cys covering, also confirmed by SR-XPS data. Moreover, AgNPs ecosafety was confirmed by ecotoxicity tests which showed no effects on algal growth of both freshwater *R. subcapitata* and marine *P. tricornutum* algae in the range of tested concentrations (10–500 mg/L). Our results further support the hypothesis that the Cit/L-cys coating of AgNPs prevent dissolution of Ag^{+} ions in both fresh and saltwater. These results open new ways for AgNPs sensing applications in environmental tests on more complex biological systems, up to tests on real environmental aquatic scenarios.

Supplementary Materials: The following are available online at http://www.mdpi.com/2079-4991/9/10/1353/s1, Figure S1. TEM images of AgNPs; Figure S2. L-Cys FTIR spectrum; Figure S3. Uv-Vis spectra of of AgNPs in presence of 2.5 ppm of metal ions (As^{3+}, As^{5+}, Ca^{2+}, Co^{2+}, Mg^{2+}, Nd^{3+}, Ni^{2+}, Zn^{2+}, Pb^{2+}) in water; all references curves (AgNPs alone) are in black line, while red curves represent the interaction between AgNPs and metal ions; Table S1. Table XPS. BE, FWHM, Atomic Ratio values and proposed assignments for all measured core-level signals title.

Author Contributions: I.V. designed and made the chemical experiments and synthesis; P.P. and L.B. performed sensing tests; I.C., G.P., C.F., and A.B. carried out the AgNPs biological eValuation and TEM, ICP-MS and DLS measurements; G.I. and L.T. performed measurements and the data analysis by ATR-FTIR; G.I. and V.S. performed measurements and the data analysis, by NEXAFS; C.B. performed measurements and the data analysis, by SR-XPS; S.F. provided technical support in SR-XPS measurements. All authors contributed to the paper writing.

Funding: This research received no external funding.

Acknowledgments: The Grant of Excellence Departments, MIUR (ARTICOLO 1, COMMI 314–337 LEGGE 232/2016), is gratefully acknowledged by authors of Roma Tre University. This research was partially funded by Regione Lazio, through Progetto di ricerca 85-2017-15125, according to L.R.13/08. We gratefully acknowledge the assistance of our colleagues at Elettra for providing good quality synchrotron light. CERIC-ERIC consortium, Grant Agency of Charles University (GAUK project No. 1054217) and Czech Ministry of Education (LM2015057) are acknowledged for the access to experimental facility and financial support.

Conflicts of Interest: The authors report no conflict of interest in this work.

References

1. Zhang, Z.; Shen, W.; Xue, J.; Liu, Y.; Liu, Y.; Yan, P.; Liu, J.; Tang, J. Recent advances in synthetic methods and applications of silver nanostructures. *Nanoscale Res. Lett.* **2018**, *13*, 54. [CrossRef] [PubMed]
2. Lohse, S.E.; Murphy, C.J. Applications of colloidal inorganic nanoparticles: From medicine to energy. *J. Am. Chem. Soc.* **2012**, *134*, 15607–15620. [CrossRef] [PubMed]
3. Venditti, I. Gold nanoparticles in photonic crystals applications: A review. *Materials* **2017**, *10*, 97. [CrossRef] [PubMed]
4. Giner-Casares, J.J.; Henriksen-Lacey, M.; Coronado-Puchau, M.; Liz-Marzán, L.M. Inorganic nanoparticles for biomedicine: Where materials scientists meet medical research. *Mater. Today* **2016**, *19*, 19–28. [CrossRef]
5. Naponiello, G.; Venditti, I.; Zardetto, V.; Saccone, D.; Di Carlo, A.; Fratoddi, I.; Barolo, C.; Dini, D. Photoelectrochemical characterization of squaraine-sensitized nickel oxide cathodes deposited via screen-printing for p-type dye-sensitized solar cells. *Appl. Surf. Sci.* **2015**, *356*, 911–920. [CrossRef]
6. Dadashi, S.; Poursalehi, R.; Delavari, H. Optical and structural properties of oxidation resistant colloidal bismuth/gold nanocomposite: An efficient nanoparticles based contrast agent for X-ray computed tomography. *J. Mol. Liq.* **2018**, *254*, 12–19. [CrossRef]
7. de Angelis, R.; Venditti, I.; Fratoddi, I.; de Matteis, F.; Prosposito, P.; Cacciotti, I.; D'Amico, L.; Nanni, F.; Yadav, A.; Casalboni, M.; et al. From nanospheres to microribbons: Self-assembled Eosin Y doped PMMA nanoparticles as photonic crystals. *J. Colloid Interface Sci.* **2014**, *414*, 24–32. [CrossRef] [PubMed]
8. Lin, C.-Y.; Yu, C.-J.; Lin, Y.-H.; Tseng, W.-L. Colorimetric sensing of silver(I) and mercury(II) ions based on an assembly of tween 20-stabilized gold nanoparticles. *Anal. Chem.* **2010**, *82*, 6830–6837. [CrossRef] [PubMed]
9. Venditti, I. Morphologies and functionalities of polymeric nanocarriers as chemical tools for drug delivery: A review. *J. King Saud Univ. Sci.* **2019**, *31*, 398–411. [CrossRef]
10. Chai, F.; Wang, C.; Wang, T.; Ma, Z.; Su, Z. L-Cysteine functionalized gold nanoparticles for the colorimetric detection of Hg^{2+} induced by ultraviolet light. *Nanotechnology* **2010**, *21*, 025501. [CrossRef] [PubMed]
11. Ciotta, E.; Paoloni, S.; Richetta, M.; Prosposito, P.; Tagliatesta, P.; Lorecchio, C.; Venditti, I.; Fratoddi, I.; Casciardi, S.; Pizzoferrato, R. Sensitivity to heavy-metal ions of unfolded fullerene quantum dots. *Sensors* **2017**, *17*, 2614. [CrossRef] [PubMed]

12. Chen, K.-J.; Chen, H.-L.; Tang, C.-C.; Wu, H.-H.; Jan, J.-S. Synthesis of silica/polypeptide hybrid nanomaterials and mesoporous silica by molecular replication of sheet-like polypeptide complexes through biomimetic mineralization. *J. Colloid Interface Sci.* **2019**, *542*, 243–252. [CrossRef] [PubMed]
13. Burratti, L.; Ciotta, E.; Bolli, E.; Kaciulis, S.; Casalboni, M.; De Matteis, F.; Garzón-Manjón, A.; Scheu, C.; Pizzoferrato, R.; Prosposito, P. Fluorescence enhancement induced by the interaction of silver nanoclusters with lead ions in water. *Colloid Surf. Sci. A* **2019**, *579*, 123634. [CrossRef]
14. Corsi, P.; Venditti, I.; Battocchio, C.; Meneghini, C.; Bruni, F.; Prosposito, P.; Mochi, F.; Capone, B. Designing an optimal ion adsorber on the nanoscale: A simple theoretical model for the unusual nucleation of AgNPs/Co^{2+}–Ni^{2+} binary mixtures. *J. Phys. Chem. C* **2019**, *123*, 3855–3860. [CrossRef]
15. Jeevanandam, J.; Barhoum, A.; Chan, Y.S.; Dufresne, A.; Danquah, M.K. Review on nanoparticles and nanostructured materials: History, sources, toxicity and regulations. *Beilstein J. Nanotechnol.* **2018**, *9*, 1050–1074. [CrossRef] [PubMed]
16. Rossi, S.; Donadio, S.; Fontana, L.; Porcaro, F.; Battocchio, C.; Venditti, I.; Bracci, L.; Fratoddi, I. Negatively charged gold nanoparticles as dexamethasone carrier: Stability and citotoxic activity. *RCS Adv.* **2016**, *6*, 99016–99022.
17. Burdușel, A.-C.; Gherasim, O.; Grumezescu, A.M.; Mogoantă, L.; Ficai, A.; Andronescu, E. Biomedical applications of silver nanoparticles: An up-to-date overview. *Nanomaterials* **2018**, *8*, 681. [CrossRef] [PubMed]
18. Liu, G.; Gao, H.; Li, K.; Xiang, J.; Lan, T.; Zhang, Z. Fabrication of silver nanoparticle sponge leather with durable antibacterial property. *J. Colloid Interface Sci.* **2018**, *514*, 338–348. [CrossRef] [PubMed]
19. Nguyen, N.D.; Nguyen, T.V.; Chu, A.D.; Tran, H.V.; Tran, L.T.; Huyn, C.D. A label-free colorimetric sensor based on silver nanoparticles directed to hydrogen peroxide and glucose. *Arab. J. Chem.* **2018**, *11*, 1134–1143. [CrossRef]
20. Rinaldi, F.; del Favero, E.; Moeller, J.; Hanieh, P.N.; Passeri, D.; Rossi, M.; Angeloni, L.; Venditti, I.; Marianecci, C.; Carafa, M.; et al. Hydrophilic silver nanoparticles loaded into niosomes: Physical-chemical characterization in view of biological applications. *Nanomaterials* **2019**, *9*, 1177. [CrossRef]
21. Bindhu, M.R.; Umadevi, M. Silver and gold nanoparticles for sensor and antibacterial applications. *Spectrochim. Acta Part A Mol. Biomol. Spectrosc.* **2014**, *128*, 37–45. [CrossRef] [PubMed]
22. Porcaro, F.; Carlini, L.; Ugolini, A.; Visaggio, D.; Luisetto, I.; Visca, P.; Fratoddi, I.; Venditti, I.; Simonelli, L.; Marini, C.; et al. Synthesis and structural characterization of silver nanoparticles stabilized with 3-mercapto-1-propansulfonate and 1-thioglucose mixed thiols for antibacterial applications. *Materials* **2016**, *9*, 1028. [CrossRef] [PubMed]
23. Katok, K.V.; DWhitby, R.L.; Fukuda, T.; Maekawa, T.; Bezverkhyy, I.; Mikhalovsky, S.V.; Cundy, A.B. Hyperstoichiometric interaction between silver and mercury at the nanoscale. *Angew. Chem. Int. Ed.* **2012**, *51*, 2632–2635. [CrossRef] [PubMed]
24. Prosposito, P.; Mochi, F.; Ciotta, E.; Casalboni, M.; Venditti, I.; Fontana, L.; Testa, G.; Fratoddi, I. Hydrophilic silver nanoparticles with tunable optical properties: Application for the detection of heavy metals in water. *Beilstein J. Nanotechnol.* **2016**, *7*, 1654–1661. [CrossRef] [PubMed]
25. Bothra, S.; Solanki, J.N.; Sahoo, S.K. Functionalized silver nanoparticles as chemosensor for pH, Hg^{2+} and Fe^{3+} in aqueous medium. *Sens. Actuators B Chem.* **2013**, *188*, 937–943. [CrossRef]
26. Brandon, M.P.; Ledwith, D.M.; Kelly, J.M. Preparation of saline-stable, silica-coated triangular silver nanoplates of use for optical sensing. *J. Colloid Interface Sci.* **2014**, *415*, 77–84. [CrossRef] [PubMed]
27. Mochi, F.; Burratti, L.; Fratoddi, I.; Venditti, I.; Battocchio, C.; Carlini, L.; Iucci, G.; Casalboni, M.; De Matteis, F.; Casciardi, S.; et al. Interaction of colloidal silver nanoparticles with Co^{2+} and Ni^{2+} in water for sensing application. *Nanomaterials* **2018**, *8*, 488. [CrossRef]
28. Contino, A.; Maccarrone, G.; Zimbone, M.; Reitano, R.; Musumeci, P.; Calcagno, L.; Oliveri, I.P. Tyrosine capped silver nanoparticles: A new fluorescent sensor for the quantitative determination of copper(II) and cobalt(II) ions. *J. Colloid Interface Sci.* **2016**, *462*, 216–222. [CrossRef]
29. Chen, N.; Zhang, Y.; Liu, H.; Wu, X.; Li, Y.; Miao, L.; Shen, Z.; Wu, A. High-performance colorimetric detection of Hg^{2+} based on triangular silver nanoprisms. *ACS Sens.* **2016**, *1*, 521–527. [CrossRef]
30. Ghosh, S.; Maji, S.; Mondal, A. Study of selective sensing of Hg^{2+} ions by green synthesized silver nanoparticles suppressing the effect of Fe^{3+} ions. *Colloids Surf. A* **2018**, *555*, 324–331. [CrossRef]

31. Zheng, W.; Liang, L.; Gu, B. Mercury reduction and oxidation by reduced natural organic matter in anoxic environments. *Environ. Sci. Technol.* **2012**, *46*, 292–299. [CrossRef] [PubMed]
32. Annadhasan, M.; Rajendiran, N. Highly selective and sensitive colorimetric detection of Hg(II) ions using green synthesized silver nanoparticles. *RSC Adv.* **2015**, *5*, 94513–94518. [CrossRef]
33. Jarujamrus, P.; Amatatongchai, M.; Thima, A.; Khongrangdee, T.; Mongkontong, C. Selective colorimetric sensors based on the monitoring of an unmodified silver nanoparticles (AgNPs) reduction for a simple and rapid determination of mercury. *Spectrochim. Acta Part A* **2015**, *142*, 86–93. [CrossRef] [PubMed]
34. Stobieck, M.; Coopersmith, K.; Hepel, M. Resonance elastic light scattering (RELS) spectroscopy of fast non-Langmuirian ligand-exchange in glutathione-induced gold nanoparticle assembly. *J. Colloid Interface Sci.* **2010**, *350*, 168–177. [CrossRef] [PubMed]
35. Stobieck, M.; Hepel, M. Rapid functionalization of metal nanoparticles by moderator-tunable ligand-exchange process for biosensor designs. *Sens. Actuators B Chem.* **2010**, *149*, 373–380. [CrossRef]
36. Duan, J.; Yin, H.; Wei, R.; Wang, W. Facile colorimetric detection of Hg^{2+} based on antiaggregation of silver nanoparticles. *Biosens. Bioelectron.* **2014**, *57*, 139–142. [CrossRef] [PubMed]
37. Zhan, L.; Yang, T.; Zhen, S.J.; Huang, C.Z. Cytosine triphosphate-capped silver nanoparticles as a platform for visual and colorimetric determination of mercury(II) and chromium(III). *Microchim. Acta* **2017**, *184*, 3171–3178. [CrossRef]
38. Farhadi, K.; Forough, M.; Molaei, R.; Hajizadeh, S.; Rafipour, A. Highly selective Hg^{2+} colorimetric sensor using green synthesized and unmodified silver nanoparticles. *Sens. Actuators B Chem.* **2012**, *161*, 880–885. [CrossRef]
39. Ravi, S.S.; Christena, L.R.; Subramanian, N.S.; Anthony, S.P. Green synthesized silver nanoparticles for selective colorimetric sensing of Hg^{2+} in aqueous solution at wide pH range. *Analyst* **2013**, *138*, 4370–4377. [CrossRef]
40. Kumar, V.; Singh, D.K.; Mohan, S.; Bano, D.; Gundampati, R.K.; Hasan, S.H. Green synthesis of silver nanoparticle for the selective and sensitive colorimetric detection of mercury (II) ion. *J. Photochem. Photobiol. B Biol.* **2017**, *168*, 67–77. [CrossRef]
41. Otto, M.; Floyd, M.; Bajpai, S. Nanotechnology for site remediation. *Remediat. J.* **2008**, *19*, 99–108. [CrossRef]
42. Corsi, I.; Cherr, G.N.; Lenihan, H.S.; Labille, J.; Hasselov, M.; Canesi, L.; Dondero, F.; Frenzilli, G.; Hristozov, D.; Puntes, V.; et al. Common strategies and technologies for the ecosafety assessment and design of nanomaterials entering the marine environment. *ACS Nano* **2014**, *8*, 9694–9709. [CrossRef] [PubMed]
43. Corsi, I.; Winther-Nielsen, M.; Sethi, R.; Punta, C.; Della Torre, C.; Libralato, G.; Lofrano, G.; Sabatini, L.; Aiello, M.; Fiordi, L. Ecofriendly nanotechnologies and nanomaterials for environmental applications: Key issue and consensus recommendations for sustainable and ecosafe nanoremediation. *Ecotoxicol. Environ. Saf.* **2018**, *154*, 237–244. [CrossRef] [PubMed]
44. Ivask, A.; Kurvet, I.; Kasemets, K.; Blinova, I.; Aruoja, V.; Suppi, S.; Vija, H.; Käkinen, A.; Titma, T.; Heinlaan, M. Size-dependent toxicity of silver nanoparticles to bacteria, yeast, algae, crustaceans and mammalian cells in vitro. *PLoS ONE* **2014**, *9*, e102108. [CrossRef] [PubMed]
45. Navarro, E.; Piccapietra, F.; Wagner, B.; Marconi, F.; Kaegi, R.; Odzak, N.; Sigg, L.; Behra, R. Toxicity of silver nanoparticles to *Chlamydomonas reinhardtii*. *Environ. Sci. Technol.* **2008**, *42*, 8959–8964. [CrossRef] [PubMed]
46. He, D.; Dorantes-Aranda, J.J.; Waite, T.D. Silver nanoparticle–algae interactions: Oxidative dissolution, reactive oxygen species generation and synergistic toxic effects. *Environ. Sci. Technol.* **2012**, *46*, 8731–8738. [CrossRef]
47. Navarro, E.; Wagner, B.; Odzak, N.; Sigg, L.; Behra, R. Effects of differently coated silver nanoparticles on the photosynthesis of *Chlamydomonas reinhardtii*. *Environ. Sci. Technol.* **2015**, *49*, 8041–8047. [CrossRef]
48. Sendra, M.; Yeste, M.; Gatica, J.; Moreno-Garrido, I.; Blasco, J. Direct and indirect effects of silver nanoparticles on freshwater and marine microalgae (*Chlamydomonas reinhardtii* and *Phaeodactylum tricornutum*). *Chemosphere* **2017**, *179*, 279–289. [CrossRef]
49. Becaro, A.A.; Jonsson, C.M.; Puti, F.C.; Siqueira, M.C.; Mattoso, L.H.; Correa, D.S.; Ferreira, M.D. Toxicity of PVA-stabilized silver nanoparticles to algae and microcrustaceans. *Environ. Nanotechnol. Monit. Manag.* **2015**, *3*, 22–29. [CrossRef]
50. Ale, A.; Liberatori, G.; Vannuccini, M.L.; Bergami, E.; Ancora, S.; Mariotti, G.; Bianchi, N.; Galdopórpora, J.M.; Desimone, M.F.; Cazenave, J.; et al. Exposure to a nanosilver-enabled consumer product results in similar

accumulation and toxicity of silver nanoparticles in the marine mussel *Mytilus galloprovincialis*. *Aquat. Toxicol.* **2019**, *211*, 46–56. [CrossRef]
51. Ribeiro, F.; Gallego-Urrea, J.A.; Jurkschat, K.; Crossley, A.; Hassellöv, M.; Taylor, C.; Soares, A.M.; Loureiro, S. Silver nanoparticles and silver nitrate induce high toxicity to *Pseudokirchneriella subcapitata*, *Daphnia magna* and *Danio rerio*. *Sci. Total Environ.* **2014**, *466*, 232–241. [CrossRef] [PubMed]
52. Kittler, S.; Greulich, C.; Diendorf, J.; Koller, M.; Epple, M. Toxicity of silver nanoparticles increases during storage because of slow dissolution under release of silver ions. *Chem. Mater.* **2010**, *22*, 4548–4554. [CrossRef]
53. Gliga, A.R.; Skoglund, S.; Wallinder, I.O.; Fadeel, B.; Karlsson, H.L. Size-dependent cytotoxicity of silver nanoparticles in human lung cells: The role of cellular uptake, agglomeration and Ag release. *Part. Fibre Toxicol.* **2014**, *11*, 11. [CrossRef] [PubMed]
54. Miao, A.-J.; Schwehr, K.A.; Xu, C.; Zhang, S.-J.; Luo, Z.; Quigg, A.; Santschi, P.H. The algal toxicity of silver engineered nanoparticles and detoxification by exopolymeric substances. *Environ. Pollut.* **2009**, *157*, 3034–3041. [CrossRef] [PubMed]
55. D'Amato, R.; Medei, L.; Venditti, I.; Russo, M.V.; Falconieri, M. Chemical synthesis of polyphenylacetylene nanospheres with controlled dimensions for photonic crystals. *Mater. Sci. Eng. C* **2003**, *23*, 861–865. [CrossRef]
56. Venditti, I.; D'Amato, R.; Russo, M.V.; Falconieri, M. Synthesis of conjugated polymeric nanobeads for photonic bandgap materials. *Sens. Actuetors B Chem.* **2007**, *126*, 35–40. [CrossRef]
57. Moulder, J.F.; Stickle, W.F.; Sobol, P.E.; Bomben, K.D. *Handbook of X-ray Photoelectron Spectroscopy*; Prairie, E., Ed.; Physical Electronics Inc.: Chanhassen, MN, USA, 1996.
58. Shirley, D.A. High-resolution X-ray photoemission spectrum of the valence bands of gold. *Phys. Rev. B* **1972**, *12*, 4709. [CrossRef]
59. Beamson, G.; Briggs, D. High resolution XPS of organic polymers. In *The Scienta ESCA 300 Database*; John Wiley & Sons: Chichester, UK, 1992.
60. Venditti, I.; Testa, G.; Sciubba, F.; Carlini, L.; Secchi, V.; Krause, S.; Meneghini, C.; Mobilio, S.; Battocchio, C.; Fratoddi, I. Hydrophilic metal nanoparticles functionalized by 2-diethylaminoethane thiol: A close look on the metal-ligand interaction and interface chemical structure. *J. Phys. Chem. C* **2017**, *121*, 8002–8013. [CrossRef]
61. Majzik, A.; Fülöp, L.; Csapó, E.; Bogár, F.; Martinek, T.; Penke, B.; Bíró, G.; Dékány, I. Functionalization of gold nanoparticles with amino acid, β-amyloid peptides and fragment. *Colloids Surf. B Biointerfaces* **2010**, *81*, 235–241. [CrossRef]
62. International Organization for Standardization. *Water Quality—Marine Algal Growth Inhibition Test with Skeletonema costatum and Phaeodactylum tricornutum*; ISO: Geneva, Switzerland, 2006; p. 12.
63. Organisation for Economic Co-operation and Development. *Test No. 201: Freshwater Alga and Cyanobacteria, Growth Inhibition Test, OECD Guidelines for the Testing of Chemicals, Section 2*; OECD Publishing: Paris, France, 2011. [CrossRef]
64. Leal, P.P.; Hurd, C.L.; Sander, S.G.; Armstrong, E.; Roleda, M.Y. Copper ecotoxicology of marine algae: A methodological appraisal. *Chem. Ecol.* **2016**, *32*, 786–800. [CrossRef]
65. Resgalla Jr, C.; Poleza, F.; Souza, R.; Máximo, M.; Radetski, C. eValuation of effectiveness of EDTA and sodium thiosulfate in removing metal toxicity toward sea urchin embryo-larval applying the TIE. *Chemosphere* **2012**, *89*, 102–107. [CrossRef] [PubMed]
66. Silverstein, R.M.; Webster, F.X.; Bryce, D.L. *Spectrometric Identification of Organic Compounds*; Wiley: Hoboken, NJ, USA, 2014.
67. Frost, M.S.; Dempsey, M.J.; Whitehead, D.E. The response of citrate functionalised gold and silver nanoparticles to the addition of heavy metal ions. *Colloids Surf. A* **2017**, *518*, 15–24. [CrossRef]
68. Li, S.; Liu, P.; Wang, Q. Study on the effect of surface modifier on selfaggregation behavior of Ag nanoparticle. *Appl. Surf. Sci.* **2012**, *263*, 613–618. [CrossRef]
69. Shi, J.; Sun, X.; Zou, X.; Zhang, H. Amino acid-dependent transformations of citrate-coated silver nanoparticles: Impact on morphology, stability and toxicity. *Toxicol. Lett.* **2014**, *229*, 17–24. [CrossRef] [PubMed]
70. Zubavichus, Y.; Shaporenko, A.; Grunze, M.; Zharnikov, M. NEXAFS spectroscopy of biological molecules: From aminoacids to functional proteins. *Nucl. Instrum. Methods Phys. Res. A* **2009**, *603*, 111–114. [CrossRef]
71. Stohr, J. *NEXAFS Spectroscopy*; Gomer, C., Ed.; Springer Science & Business Media: Berlin, Germany, 1991.

72. National Institute of Standards and Technology. *NIST X-ray Photoelectron Spectroscopy Database, Version 4.1*; National Institute of Standards and Technology: Gaithersburg, MD, USA, 2012. Available online: http://srdata.nist.gov/xps/ (accessed on 20 September 2019).
73. Gunsolus, I.L.; Mousavi, M.P.; Hussein, K.; Bühlmann, P.; Haynes, C.L. Effects of humic and fulvic acids on silver nanoparticle stability, dissolution, and toxicity. *Environ. Sci. Technol.* **2015**, *49*, 8078–8086. [CrossRef] [PubMed]
74. Forough, M.; Farhadi, K. Biological and green synthesis of silver nanoparticles. *Turk. J. Eng. Environ. Sci.* **2010**, *34*, 281–287.
75. Ma, Y.; Pang, Y.; Liu, F.; Xu, H.; Shen, X. Microwave-assisted ultrafast synthesis of silver nanoparticles for detection of Hg^{2+}. *Spectrochim. Acta Part A* **2016**, *153*, 206–211. [CrossRef]
76. Buduru, P.; Reddy, B.S.R.; Naidu, N.V.S. Functionalization of silver nanoparticles with glutamine and histidine for simple and selective detection of Hg^{2+} ion in water samples. *Sens. Actuators B Chem.* **2017**, *244*, 972–982. [CrossRef]
77. Alam, A.; Ravindran, A.; Chandran, P.; Khan, S.S. Highly selective colorimetric detection and estimation of Hg^{2+} at nano-molar concentration by silver nanoparticles in the presence of glutathione. *Spectrochim. Acta Part A* **2015**, *137*, 503–508. [CrossRef]
78. Carlini, L.; Fasolato, C.; Postorino, P.; Fratoddi, I.; Venditti, I.; Testa, G.; Battocchio, C. Comparison between silver and gold nanoparticles stabilized with negatively charged hydrophilic thiols: SR-XPS and SERS as probes for structural differences and similarities. *Colloids Surf. A* **2017**, *532*, 183–188. [CrossRef]
79. Sun, T.S.; Buchner, S.P.; Byer, N.E. Oxide and interface properties of anodic films on $Hg_{1-x}Cd_xTe$. *J. Vac. Sci. Technol.* **1980**, *17*, 1067–1073. [CrossRef]
80. Sharma, V.K.; Siskova, K.M.; Zboril, R.; Gardea-Torresdey, J.L. Organic-coated silver nanoparticles in biological and environmental conditions: Fate, stability and toxicity. *Adv. Colloid Interface Sci.* **2014**, *204*, 15–34. [CrossRef]
81. Sikder, M.; Lead, J.R.; Chandler, G.T.; Baalousha, M. A rapid approach for measuring silver nanoparticle concentration and dissolution in seawater by UV–Vis. *Sci. Total Environ.* **2018**, *618*, 597–607. [CrossRef] [PubMed]
82. Luoma, S.N. Silver nanotechnologies and the environment: Old problems or new challenges? *Proj. Emerg. Nanotechnol. Rep.* **2008**, *15*, 12–13.
83. Liu, J.; Sonshine, D.A.; Shervani, S.; Hurt, R.H. Controlled release of biologically active silver from nanosilver surfaces. *ACS Nano* **2010**, *4*, 6903–6913. [CrossRef]
84. Gondikas, A.P.; Morris, A.; Reinsch, B.C.; Marinakos, S.M.; Lowry, G.V.; Hsu-Kim, H. Cysteine-induced modifications of zero-valent silver nanomaterials: Implications for particle surface chemistry, aggregation, dissolution, and silver speciation. *Environ. Sci. Technol.* **2012**, *46*, 7037–7045. [CrossRef]
85. Mu, L.; Gao, Y.; Hu, X. L-Cysteine: A biocompatible, breathable and beneficial coating for graphene oxide. *Biomaterials* **2015**, *52*, 301–311. [CrossRef]

nanomaterials

MDPI

Article

Hydrophilic Silver Nanoparticles Loaded into Niosomes: Physical–Chemical Characterization in View of Biological Applications

Federica Rinaldi [1], Elena del Favero [2], Johannes Moeller [3,†], Patrizia Nadia Hanieh [4], Daniele Passeri [5], Marco Rossi [5], Livia Angeloni [5], Iole Venditti [6,*], Carlotta Marianecci [4,*], Maria Carafa [4,‡] and Ilaria Fratoddi [7,‡]

1 Center for Life Nano Science@Sapienza, Istituto Italiano di Tecnologia (ITT), 00161 Rome, Italy
2 Department of Medical Biotechnology and Translational Medicine, University of Milan, 20122 Milano, Italy
3 ESRF Grenoble Fr, 38043 Grenoble, France
4 Department of Drug Chemistry and Technologies, Sapienza University, 00185 Rome, Italy
5 Department of Basic and Applied Sciences for Engineering, Sapienza University, 00185 Rome, Italy
6 Department of Sciences, Roma Tre University, 00146 Rome, Italy
7 Department of Chemistry, Sapienza University, 00185 Rome, Italy
* Correspondence: iole.venditti@uniroma3.it (I.V.); carlotta.marianecci@uniroma1.it (C.M.); Tel.: +39-06-57333388 (I.V.); +39-06-49913970 (C.M.)
† Current affiliation: European XFEL, Holzkoppel 4, 22869 Schenefeld, Germany.
‡ Co-last authors.

Received: 8 July 2019; Accepted: 15 August 2019; Published: 17 August 2019

Abstract: Silver nanoparticles (AgNPs) are widely used as antibacterial agents and anticancer drugs, but often their low stability limits their mass production and broad applications. The use of niosomes as a carrier to protect and envelop AgNPs gives a new perspective to solve these problems. In this study, AgNPs were functionalized with sodium 3-mercapto-1-propanesulfonate (3MPS) to induce hydrophilic behavior, improving loading in Tween 20 and Span 20 niosomes (NioTw20 and NioSp20, respectively). Entrapment efficiency was evaluated by UV analyses and is around 1–4%. Dimensions were investigated by means of dynamic light scattering (DLS) ($<2R_H> = 140 \pm 4$ nm and $<2R_H> = 251 \pm 1$ nm respectively for NioTw20 + AgNPs and NioSp20 + AgNPs) and were compared with those by atomic force microscopy (AFM) and small angle X ray scattering (SAXS) analyses. Stability was assessed in water up to 90 days, and both in bovine serum and human serum for up to 8 h. In order to characterize the local structure of niosomes, SAXS measurements have been performed on Tween 20 and Span 20 empty niosomes and loaded with AgNPs. The release profiles of hydrophilic probe calcein and lipophilic probe Nile Red were performed in HEPES buffer and in human serum. All these features contribute to conclude that the two systems, NioTw20 + AgNPs and NioSp20 + AgNPs, are suitable and promising in the field of biological applications.

Keywords: niosomes; silver nanoparticles; liposomes; plasmonic materials; drug delivery; nanocarriers

1. Introduction

It is well known that the body barrier to external pathogenic attacks is represented by the skin, which prevents microbial invasion, so every damage or wound can provide an environment for microbial growth, leading to infection and prolonged wound healing [1–6]. The efficacy of antibiotics is superior to that of other drugs; thus, antibiotics are widely used [7]. However, the antibiotic resistance is currently a health emergency, so drug-resistant bacterial infections are becoming more common with a consequent increase in public spending [8,9]. About 60–70% of the existing antibiotics are not active against intracellular infections due to their low intracellular retention as a result of their poor permeability. Nanomaterials represent an attractive solution for the hydrophilicity barrier, because they can habitually penetrate cells, and increase their intracellular activity [10,11]. Small-size nanoparticles have the advantage of being characterized by a larger contact surface, which is able to enhance their penetration and therapeutic effects [12–15].

Among others nanomaterials, a good candidate is represented by silver nanoparticles (AgNPs) that can be easy functionalized [16,17], inducing hydrophilic behavior [18,19], antibacterial properties [13,20], and reduced inflammatory response [21], with low resistance phenomena in Gram-positive bacteria and Gram-negative bacteria [22–24]. The dimension and shape of AgNPs have a strong influence on antibacterial activity. In fact, a higher surface/volume ratio produces the higher rate of silver ions release [25]. AgNPs' effects have been studied against the multidrug-resistant bacteria such as *P. aeruginosa, E. coli, Streptococcus pyogenes, S. aureus, Klebsiella pneumoniae, Salmonella* species, and *Enterococcus species* [26,27]. In these papers, the bactericidal action is mainly due to the inhibition of cell wall synthesis, nucleic acid synthesis, and protein synthesis mediated by the 30S ribosomal subunit. The strong bactericidal effect of AgNPs against the multidrug-resistant bacteria is mostly due to their multiple mechanisms to disrupt microbial cells [28]. Moreover, AgNPs can improve the antibiotic effects against *S. aureus* and *E. coli* [29].

Today, in medicinal practice, there are wound dressings, contraceptive devices, surgical instruments, bone prostheses, and dental implants that are coated or embedded with nanosilver [30–33]. Moreover, in the last decade, the research field of AgNPs has moved to the possibility of their use as an anticancer drug, due to their inherent cytotoxic effect on cancer cells [34].

However, the instability of silver nanoparticles limits their industrial application in several cases, and most of the methods to prepare AgNPs cause environmental pollution and low production efficiency. To overcome this problem, silver nanoparticles are usually loaded onto carriers [35].

Moreover, despite AgNPs having multiple mechanisms for antibacterial effects, recent studies showed bacterial resistance to them: the resistance evolves without any genetic changes; only phenotypic change is needed to reduce the nanoparticles' colloidal stability and thus eliminate their antibacterial activity [36,37].

As a response to this problem, hybrids/composites with AgNPs dispersed on carriers or supports have been studied to enhance antibacterial activity compared with sole AgNPs: it is of significance to seek the optimal choice of carriers to combine with AgNPs in order to construct ideal antibacterial agents. Various AgNPs-based nanocomposites with different structures and morphologies have been developed up to now, such as an amorphous silica matrix dispersed with AgNPs [38], AgNPs core@silica shell [39], mesoporous silicas loaded with AgNPs [40], hollow mesoporous silica spheres with AgNPs in the cavity [41,42], fibers coated with AgNPs [43], etc. Although extensive efforts have been devoted to fabricating a lot of these AgNPs-based nanocomposites involving different carriers' structures, there are still few systematic investigations on the effects of structures on antibacterial performance [44]. The use of drug delivery systems (DDS) has been proposed to overcome important issues in the release of active pharmaceutical molecules, such as unfavorable pharmacokinetics and biodistribution with a consequent decrease of side effects.

Nanocarriers represent an innovative approach to overcome these issues [45,46]. Among other nanocarriers, such as liposomes, polymersomes, micelles, and polymer-based vesicles, the niosomes systems, non-ionic surfactant vesicles [46,47], have attracted attention from researchers because of their ability to encapsulate different kinds of drugs for the purpose of increasing their stability and efficacy. In fact, niosomes enable modulating the drug concentration loading in a range of interest for the biological applications (0.3–5.0 μg/mL for AgNPs) and to consent to drug-release control [48].

In this research study, AgNPs were synthetized using 3-mercapto-1-sodium propanesulfonate (3MPS) to induce hydrophilic behavior, improving the niosomal entrapment efficiency and reducing the bilayer destabilization. AgNPs were loaded in two different niosomes, Tween 20 and Span 20 ones, producing two different systems, namely NioTw20 + AgNPs and NioSp20 + AgNPs. A deep physical chemical characterization was carried out to obtain information on hydrophilic AgNPs and their influence on the preparation and characterization of Nio + AgNPs. Moreover, stability studies were performed in water, bovine serum, and human serum to assess their use in biological compartments. Hydrophilic and lipophilic probe release profiles were obtained in HEPES and in human serum. Both systems proved to be able to entrap AgNPs, are stable, and maintain the ability to entrap also hydrophilic or lipophilic model molecules, and so are promising systems for biotechnological applications.

2. Experimental

2.1. Materials and Methods

Silver nitrate ($AgNO_3$, 99.5%, Sigma-Aldrich, St. Louis, MO, USA) and sodium borohydride (NaBH4, 98%, Sigma-Aldrich, St. Louis, MO, USA) were used for the synthesis of the nanoparticles. 3-mercapto-1propanesulfonic acid sodium salt ($C_3H_7S_2O_3Na$, 3MPS, 98%, Sigma Aldrich, St. Louis, MO, USA) was used as a capping agent. For all of the solutions, we used deionized water (electrical conductivity less than 1 μΩ/cm at room temperature) obtained from a Millipore Milli-Q water purification system. HEPES salt (sodium 2-(4-(2-hydroxyethyl) piperazin-1-yl) ethanesulfonate), cholesterol, Sephadex G75, Pyrene, DPH (1,6-diphenyl-1,3,5-exatriene), Span 20 (sorbitan monolaurate), Tween 20 (polyoxyethylene sorbitan monolaurate), human/bovine serum, calcein, and Nile Red were purchased from Sigma-Aldrich (Milan, Italy). All the other products and reagents were of analytical grade. All of the reagents were purchased from Sigma Aldrich and were used without further purification.

2.2. Preparation of AgNPs Loaded Niosomes

The AgNPs–3MPS synthesis (see Figure S1) consisted in a wet reduction of silver nitrate to metallic silver by means of sodium borohydride in the presence of 3MPS [17,18]. Briefly, a solution of $AgNO_3$ in deionized water (0.200 g in 10 mL) was added in a flask with 3MPS water solution (0.830 g in 10 mL), and the mixture was maintained under stirring in argon atmosphere at room temperature for 10 min. Then, an $NaBH_4$ water solution (0.220 g in 10 mL) was added dropwise, under vigorous stirring. The reaction mixture was allowed to react for 2 h. The obtained black product was centrifuged with deionized water three times (20 min, 5000 rpm), and the solid obtained was characterized.

Several niosomal formulations by Tween 20 (NioTw20) or Span 20 (NioSp20) were prepared using AgNPs-3 MPS at different concentrations. Only the results obtained by selected samples in terms of the size, ζ-potential, entrapment efficiency, and stability features were reported and discussed.

AgNPs–3MPS were loaded into Span 20 and Tween 20 niosomes through a protocol already used to internalize chemicals [49]. Niosomes were prepared using the thin film hydration method [50]. Span 20 or Tween 20 (15 mM) and cholesterol (15 mM) were dissolved in organic solvent mixture (chloroform/methanol 3:1 *v*/*v*). The solvent was evaporated using a rotary evaporator (VV2000, Heidolph, Schwabach, Germany) to form a thin "film". The film was hydrated using 5 mL of AgNPs solution, which was then vortexed and sonicated at 60 °C and 18%/16% amplitude for 5 min using an ultrasonic microprobe (Vibra-Cell VCX-400, Sonics & Materials, Newtown, CT, USA). The unilamellar vesicular suspension was purified by gel filtration chromatography using Sephadex G75 (glass column of 50 × 1.2 cm) with HEPES buffer as the eluent. The purified vesicles were filtrated by using cellulose filters (pore size 1.2 μm).

2.3. Characterizations

The AgNPs water suspension has been characterized by means of UV-Vis collected using a Perkin-Elmer Lambda 19 and a Cary 100 spectrophotometer using quartz cuvettes. The dynamic light scattering (DLS) measurements on the AgNPs colloidal suspensions (0.200 mg/mL) at $T = 25.0 \pm 0.2$ °C were performed by the Malvern Zetasizer Nano ZS90 instrument (Malvern, UK), as reported in previous studies [49,51]. The ζ-potential was calculated from the measured electrophoretic mobility by means of the Smolukovsky equation [52]. UV-Vis: λ_{max} (nm) 415; DLS: <$2R_H$> 5 ± 2 nm; ζ-potential: −35 ± 2 mV.

The mean size, size distribution, and ζ-potential of empty and AgNPs-loaded niosomes were characterized by using DLS. UV-Visible spectroscopy was employed to evaluate the amount of AgNPs entrapped in niosomal formulations. The AgNPs' entrapment efficiency was expressed as encapsulation yield, i.e., the percentage of nanoparticles loaded with respect to the total amount added for the preparation. Bilayer characterization has been carried out on empty Span 20 or Tween 20 niosomes and on AgNPs-loaded ones, employing DPH and pyrene (lipophilic probes) that provided different bilayer information (fluidity, microviscosity, and polarity).

Span 20/Tween 20 (15 mM), cholesterol (15 mM), and DPH solution (2×10^{-4} M) were codissolved in chloroform/methanol, which was removed using a rotatory evaporator (VV2000, Heidolph, Schwabach, Germany), and then hydrated with HEPES buffer or AgNPs solution (0.5 mg/mL), with the same preparation methods as those mentioned above. A cellulose filter with a 450-nm cut-off (Spectrum, New Jersey, New Brunswick USA) was used to purify the DPH–niosomal formulations. Fluorescent measurements were performed (λ = 350–425 nm) using a luminescence spectrometer (LS5013, PerkinElmer) in order to obtain fluorescence anisotropy. The florescence anisotropy (r) was determined by using Equation (1) [53–55].

$$A = \frac{I_{vv} - I_{vh} \times G}{I_{vv} + 2I_{vh} \times G} \quad (1)$$

where *Ivv, Ivh, Ihv*, and *Ihh* are fluorescent intensities, and subscript *v* (vertical) and *h* (horizontal) represent the orientation of polarized light. The G factor is ratio of the sensitivity of the detection system for vertically and horizontally polarized light.

Pyrene-loaded niosomes were prepared by adding the probe (4 mM) to niosomes (Nio) components in order to obtain empty and Nio-AgNPs following the same preparation method as above. Pyrene is a florescence probe, whose monomer exhibited a spectrum characterized with five emission peaks (from I1 to I5) and excimer has only one peak (IE). The monomer and the excimer have different fluorescence signals, and the ratio between the several fluorescence intensities is directly related to the probe distribution in the bilayer. In particular, the ratio I1/I3, corresponding to the first and third vibration bands in the pyrene spectrum, is related to the polarity of the probe environment. Pyrene can form an intramolecular excimer based on the viscosity of the probe microenvironment [56], and it is estimated with the ratio IE/I3, where IE is the excimer intensity. The fluorescence signals emitted by pyrene-loaded niosome suspension were scanned (λ = 350–550 nm) using a luminescence spectrometer (LS5013, PerkinElmer) and the intensities of the excimer florescence (IE), first (I1), and third (I3) peak were recorded [57].

An atomic force microscopy (AFM) study of the morphology of the niosomes has been performed using a standard AFM setup (Dimension Icon, Bruker Inc., Milan, Italy) equipped with Si cantilevers suitable for tapping mode imaging. Samples have been prepared by depositing a droplet of solution on a clean monocrystalline Si surface and waiting until partial dehydration. AFM imaging has been performed in air and at room conditions.

Small-angle X-ray scattering (SAXS) experiments were performed at a ID02 high-brilliance beamline at the ESRF (Grenoble, France). The measured SAXS profiles report the scattered radiation intensity as a function of the momentum transfer, $q = (4\pi/\lambda) \sin(\theta/2)$, where θ is the scattering angle and λ is the X-ray wavelength (0.1 nm). Analysis was carried out to obtain information on the dimension, homogeneity, and shape of the particles in solution. The form factors of niosomes have been reconstructed as unilamellar or multilamellar closed bilayers. Details are reported in the Supporting Information.

Biological studies were also carried out in the presence of bovine or human serum to evaluate the in vitro stability. NioTw20/AgNPs and NioSp20/AgNPs, as well as empty niosomes, were diluted in fetal bovine or human serum to obtain a final serum concentration of 45% at 37 °C. The average size, polydispersity index, and ζ-potential were evaluated by means of DLS at different time points (15 min, 30 min, 60 min, and 180 min) [58].

Stability studies up to 90 days at different temperatures—room temperature (RT) and 4 °C—were carried out by DLS to asses the dimension and ζ-potential of empty niosomes and Nio-AgNPs [58].

Release studies were carried out following the release calcein from empty niosomes and Nio-AgNPs using a Fluorimetric apparatus.

In order to obtain information about the ability of Nio to release lipophilic and hydrophilic probes, also in presence of AgNPs, Nile Red (lipophilic fluorescent probe) and calcein (hydrophilic fluorescent probe) release studies were carried out.

The hydrophilic probe (calcein 10^{-2} M) was added to the film during hydration, and the excess of calcein was purified by gel filtration chromatography, as already mentioned [58].

On the contrary, due to its apolar nature, the fluorescent probe Nile Red was added to the vesicle components at a final concentration of 10^{-3} M, and it will be located in the hydrophobic bilayer.

The in vitro release experiments were carried out using dialysis tubes (molecular weight cut-off 8000 and 5.5 cm^2 diffusing area) at 37 °C in HEPES buffer (10 mM, pH 7.4). The Nile Red/calcein concentration was measured using the UV spectrophotometer/luminescence spectrometer at different time points over 1–24 h.

In vitro release experiments were performed at 37 °C, and the defined volume of vesicle dispersions was included in dialysis sacs (cut-off 8.000) with a fixed diffusing area (5.5 cm^2) adding to a niosomal formulation of 45% human serum or HEPES (in order to maintain the same probe concentration). The probe concentration was detected in the outer solution at fixed time intervals (0 h, 1 h, 2 h, 3 h, 4 h, 5 h, 6 h, 7 h, 8 h, and 24 h) by means of the Fluorometric apparatus, taking into account the dilution factor.

3. Results and Discussion

AgNPs were functionalized by 3MPS to induce high hydrophilicity and to control the shape and dimension in the range of 2 to 5 nm by means of UV-Vis and DLS measurements, as already reported [17,18]. In fact, the hydrophilicity is a crucial feature to obtain the insertion in the aqueous core of the niosomes. Even the small dimensions are a key parameter to ensure the inclusion and above all the stability of the final hybrid system, as also reported in the literature [48]. The AgNPs were loaded into niosomes, as schematized in Figure 1.

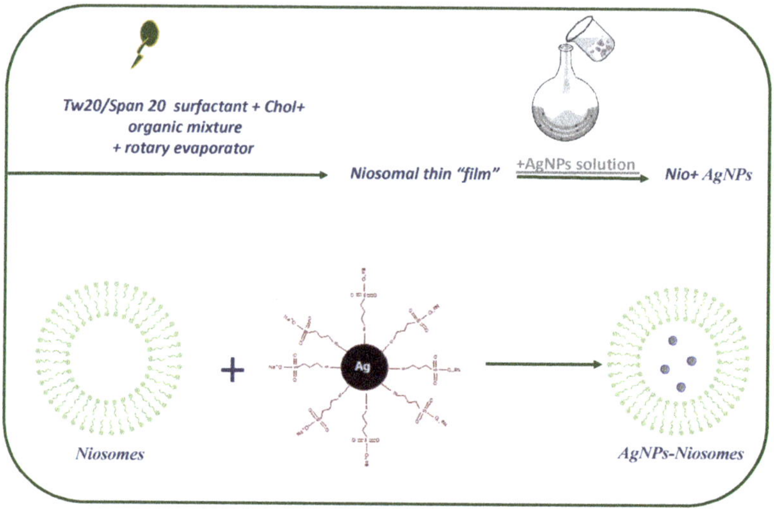

Figure 1. Preparation of niosomes hydrated with silver nanoparticles (Nio-AgNPs).

According to characterization results obtained, the best samples selected were Tween 20/Span 20 niosomes hydrated with AgNPs at the 0.5 mg/mL concentration.

The first comparison between empty niosomes and Nio-AgNPs was done by analyzing their hydrodynamic diameter, ζ-potential, and PDI (polydispersity index) by DLS. The results are shown in Table 1. In both niosomal formulations, with Tween 20 or Span 20, all the parameters analyzed by DLS are preserved after the addition of AgNPs. The empty samples based on Tween 20 and Span 20 show differences in dimensions, due to the different internal structures determined by the different surfactants employed for niosomal preparation. In particular, Span 20 niosomes, as expected [59], are bigger than Tween 20 ones, and show more negative ζ-potential.

Table 1. Hydrodynamic diameter, ζ-potential, and polydispersity index (PDI) of different niosomal formulations. NioTw20: niosomal formulations by Tween 20, NioSp20: niosomal formulations by Span 20.

Samples ID	Hydrodynamic Diameter (nm) ± SD	ζ-Potential (mV) ± SD	PDI ± SD
NioTw20	136.1 ± 2.0	−32.8 ± 0.3	0.38 ± 0.01
NioTw20 + AgNPs	140.3 ± 3.9	−33.1 ± 1.4	0.40 ± 0.01
NioSp20	230.2 ± 5.9	−42.7 ± 2.3	0.35 ± 0.01
NioSp20 + AgNPs	251.7 ± 6.0	−42.9 ± 1.2	0.40 ± 0.01

The entrapment efficiency of AgNPs in vesicular systems was evaluated by means of the calibration curve (see Figure S2), and the results are reported in Table 2.

Table 2. The entrapment efficiency in percentage of AgNPs in niosomes.

Samples ID	Entrapment Efficiency (%)
NioTw20 + AgNPs	<1
NioSp20 + AgNPs	4

The data obtained indicate that the entrapment efficiency for the two systems is not the same; Span 20 niosomes are more efficient than Tw20 niosomes, which is probably related to their different internal structures and/or capacity (aqueous volume available to host AgNPs). Moreover, these values allow assembling systems with a silver concentration in the range of interest for biological applications (0.3–5.0 μg/mL), as reported in literature [48]. The results obtained by bilayer characterization studies, such as fluidity, microviscosity, and polarity (Table 3), show no variation in the evaluated bilayer properties, thus demonstrating that no interactions occur between the niosomal double layer and AgNPs, which probably will be located inside aqueous compartments.

Table 3. Bilayer characterization results of niosomes (Nio) and Nio-AgNPs.

Samples ID	Fluidity (Anisotropy)	Microviscosity (I_E/I_3)	Polarity (I_1/I_3)
NioTw20	0.10	0.90	0.90
NioTw20 + AgNPs	0.11	0.90	0.90
NioSp20	0.10	1.01	0.94
NioSp20 + AgNPs	0.11	1.03	0.90

Morphological studies were performed on empty niosomes and Nio-AgNPs. Representative AFM images of the samples are shown in Figure 2. The morphological characterization indicates that niosomes have regular spherical shapes. Probably due to the intrinsic limitations related to sample preparation, the size of niosomes seems highly dispersed, according to PDI values by DLS analyses, the bigger particles being probably the results of the agglomeration of individual niosomes.

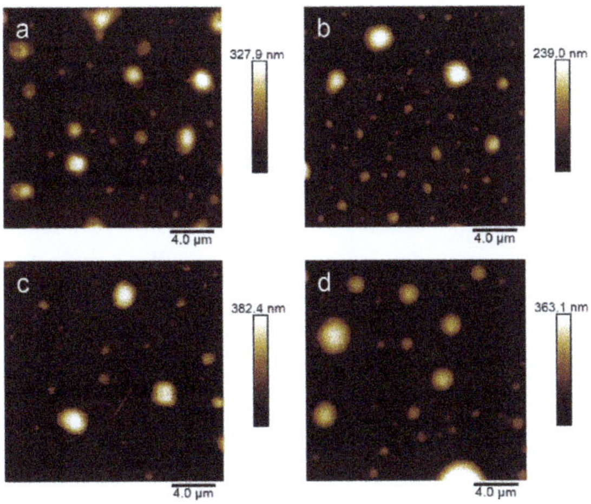

Figure 2. Atomic force microscopy (AFM) images related to: NioTw20 (**a**), NioTw20 + AgNPs (**b**), NioSp20 (**c**), and NioSp20 + AgNPs (**d**).

In the deposited sample, large amorphous particles are visible, likely resulting from the coalescence of several vesicles on the substrate surface as well as the possible partial dehydration making the vesicles lose their original spherical shape. By a visual inspection, Tween-based niosomes, i.e., NioTw20 (Figure 2a) and NioTw20 + AgNPs (Figure 2b), are smaller than the Span-based ones, i.e., NioSp20 (Figure 2c) and NioSp20 + AgNPs (Figure 2d), which is in qualitative agreement with the measured hydrodynamic diameters reported in Table 1. Conversely, the effect of the presence of AgNPs on the size cannot be appreciated in both the Tween and Span-based formulations, considering the relatively small variations evaluated by DLS and the dispersion of the size in the AFM samples.

In order to characterize the local structure of niosomes, SAXS measurements were performed on Tween 20 and Span 20 niosomes that were empty and loaded with AgNPs.

In Figure 3, the intensity spectra are reported for Tween 20-based nanoparticles (panel A) and for Span 20-based nanoparticles (panel B) in a wide q range (0.014 $nm^{-1} \leq q \leq 6$ nm^{-1}), corresponding to distances from 150 nm to the nm. Differences in the intensity profiles are clearly visible for the two systems in all the regions of the spectra. Besides the feature of the form factor of the particles in solution, a small diffraction peak at 1.84 nm^{-1}, corresponding to a characteristic distance of 3.41 nm, is well known, and stems from the presence of cholesterol crystallites in surfactants/cholesterol mixtures [60]. Crystallites could be either excluded from the bilayers, in peripheral contact, for example, or included in the bilayers, as segregated structures.

Tween 20-based particles display a niosomes shape that is characterized by a local bilayer structure with a thickness of about 6 nm, being the hydrophobic core of 2.5 nm and the two hydrophilic layers of 1.6 nm and of 2 nm, respectively. Moreover, the absence of peaks due to multilamellar organization reveals that the adopted structure is unilamellar.

In the presence of AgNPs, Tween 20 niosomes keep a local unilamellar structure, with unaffected structural features. On the other hand, in the low-q region of the spectrum, the scattered intensity increases by one order of magnitude. The form factor of the unilamellar closed particles can be modeled, replacing the internal water with a higher electron density solvent, confirming the presence of AgNPs entrapped inside the aqueous core of the niosomes.

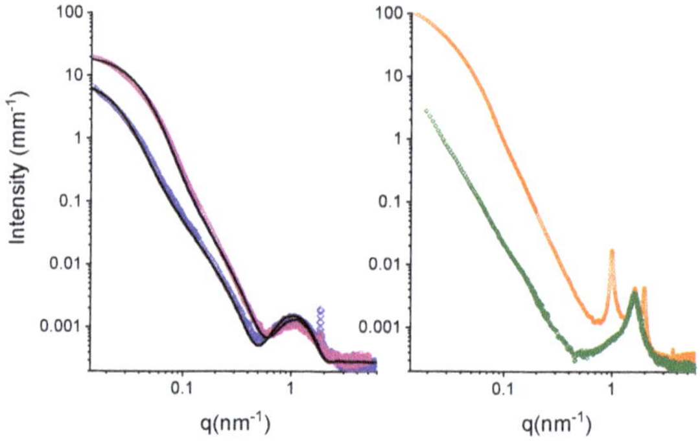

Figure 3. SAXS spectra of Tween 20 and Span 20-based niosomes. Panel A. Tween 20-based empty niosomes (blue diamonds) and Nio-AgNPs (magenta dots). Fitting curves have been obtained by modeling the particle as an internal solvent core surrounded by a surfactant closed bilayer (for details, see Supporting Information). Panel B. Span 20-based empty niosomes (green diamonds) and Nio-AgNPs (orange dots).

Span 20-based aggregates display the characteristic features of closed lamellar-type particles. Nevertheless, the local structure of Span 20-based niosomes is quite different from the Tween 20-based one. A broad intensity peak is clearly visible at q = 1.6 nm^{-1}, together with a very broad left shoulder. The position of the peak corresponds to a characteristic distance of d = 3.9 nm, which is compatible with twice the length of a Span 20 (sorbitan monolaurate) molecule. The results indicate that Span 20 niosomes are multilamellar closed particles, with a water core surrounded by a peculiar layered shell: several adjacent concentric bilayers are in close contact, heads to head, without any water penetration. The scattered intensity profile of Span 20-based loaded niosomes, as reported in Figure 3 (panel B), presents a pronounced increase in the low-q region, which is a sign of the presence of AgNps enclosed in the internal aqueous core of the niosomes. The increase is definitely higher than the one observed in Tween 20-based Nio-AgNPs, suggesting that Span 20 is more efficient in entrapping metallic NPs. On the local scale, two additional peaks at q = 1 nm^{-1} and q = 2 nm^{-1} are visible, revealing a swollen multilamellar structure with a characteristic distance of d = 6 nm, coexisting with the tight one at d = 3.9 nm. The 6-nm distance is typical for lipid multilamellar structures and is also found in Tween®20-derivatives/cholesterol niosomes [60]. The presence of AgNPs induces the partial disjunction of adjacent bilayers with increased water penetration.

Stability studies of NioTw20/NioTw20 + AgNPs and NioSp20/NioSp20 + AgNPs were performed to compare their behavior at different temperatures, as shown in Figure 4. Empty niosomes are stable both at room temperature and at 4 °C, while Nio-AgNPs showed a different behavior in terms of the dimensional increase at RT conditions, which was not confirmed at a 4 °C storage temperature. The colloidal stability at 4 °C is higher because of the reduced collision events of the dispersed particles, and hence coalescent phenomena.

The effect of the serum (human and bovine) is another important element to evaluate in order to define the interaction between vesicles and biological fluids. Experiments were performed at 37 °C evaluating the size and ζ-potential (data not shown) variations by DLS analysis up to 3 h (Figure 5). During the time interval analyzed, the same trend is observed for all the vesicles. Vesicles in 45% bovine serum do not show a dimensional increase, while in 45% human serum, the trend is different. The different protein composition of human serum is the reason for the attractive interaction between a negatively charged niosomal surface and proteins. These interactions are strong enough to observe the same populations over the three hours. This result, which has to be investigated further, is consistent with the fact that niosomal vesicles should act as an anchor for the blood proteins. Indeed, after incubation with human serum, plateau values of about 280 nm and 380 nm are reached. This time evolution suggests that the human serum composition is responsible for a faster kinetic toward the equilibrium rather than the bovine one, because of a different protein pattern [57,61].

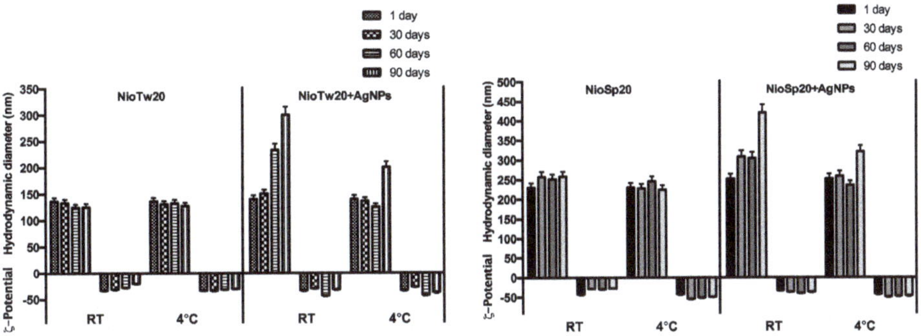

Figure 4. Stability studies of empty niosomes and Nio-AgNPs at room temperature and 4 °C.

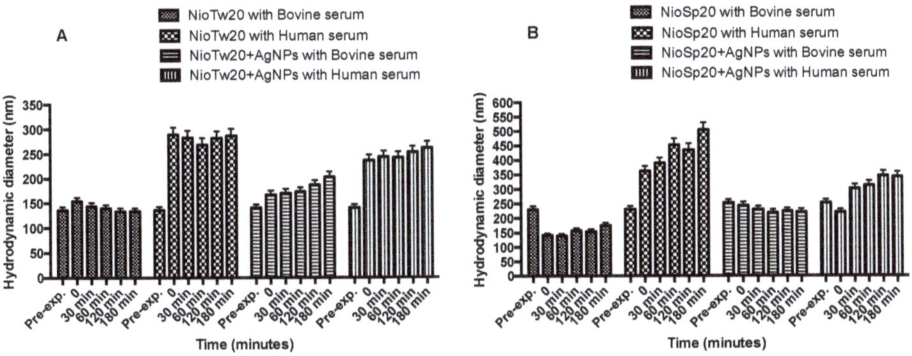

Figure 5. Niosomal biological stability at 37 °C in different media. (**A**) NioTw20/AgNPs in bovine and human serum; (**B**) NioSp20/AgNPs in bovine and human serum.

In order to confirm the ability of Nio to release lipophilic and hydrophilic probes, also in the presence of AgNPs, calcein and Nile Red release studies were carried out.

Figures 6 and 7 show the release profiles calcein and Nile Red. In each experiment, the calcein release data from empty niosomes and NioTw20 AgNPs were compared in order to evaluate the influence of the silver nanoparticles, which were entrapped in the same compartment, on the hydrophilic probe release. The calcein entrapment in both samples was comparable. Experiments were carried out at 37 °C both in HEPES and human serum. The results obtained in human serum are not reported because the calcein release was not significant (20%), which was probably due to the coating and masking effects of the serum, which make it difficult to quantify the calcein release in the external medium. In HEPES buffer, the presence of AgNPs influences the release profile of calcein only in the NioTw20 formulation. As demonstrated by SAXS analyses, only a double layer is present, so it is more susceptible to silver nanoparticles' destabilization. Calcein release by this sample is around 65% in 24 h, with respect to 30% by NioSp20/AgNPs, where the presence of different bilayers can make calcein release more difficult.

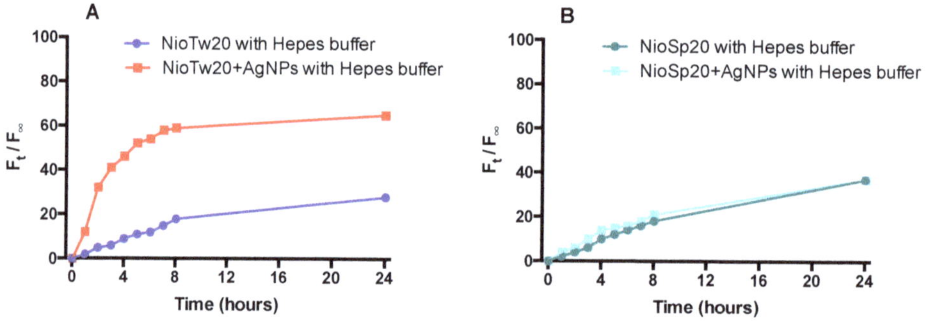

Figure 6. Calcein release studies in HEPES buffer at 37 °C from: (**A**) NioTw20/AgNPs; (**B**) NioSp20/AgNPs.

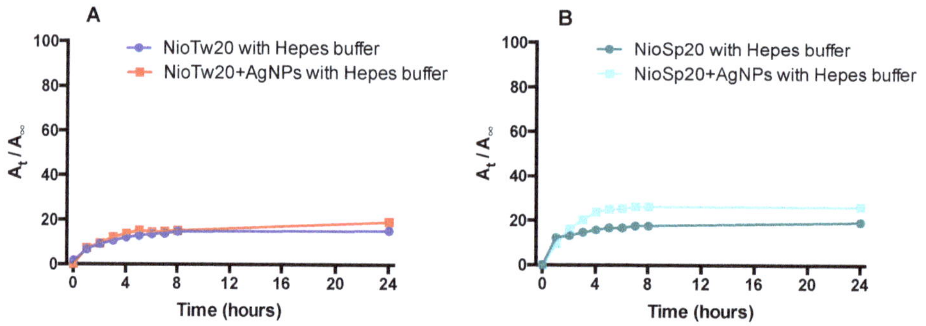

Figure 7. Nile Red release studies in HEPES buffer at 37 °C (**A**) NioTw20/AgNPs; (**B**) NioSp20/AgNPs.

On the contrary, in both samples, the Nile Red release profiles that were obtained in HEPES at 37 °C are comparable, because of the lipophilic nature of the probe, as shown in Figure 7.

The number of bilayers to cross is the limiting step for hydrophilic probe release, while it is not the limiting step for the lipophilic probe that is located in the bilayer, and will be barely released for the poor affinity with the aqueous external medium.

The morphological structure of the two different niosomal formulations and the presence or not of AgNPs, could influence the release profiles of the two probes.

4. Conclusions

In this study, hydrophilic AgNPs were loaded in two different niosomes, producing two systems, namely NioTw20 + AgNPs and NioSp20 + AgNPs. A deep physical–chemical characterization was carried out to obtain information on the influence of AgNPs on the preparation and features of niosomal formulations. First of all, the DLS studies confirm the nanosize and stability of both systems in water. Moreover, the entrapment efficiency for the two systems was investigated, and it was more efficient for Span 20 than Tw20 niosomes, which was probably related to their different internal structures. Microviscosity and polarity investigations demonstrated that no interactions occurred between the niosomal double layer and the AgNPs, which were probably located inside aqueous compartments. The SAXS data confirmed the presence of the AgNPs located inside the aqueous compartment of the two niosomal systems, and also allowed highlighting the different structures of their double layers. The morphological characterization indicates that the niosomes maintained spherical shapes. Moreover, stability was confirmed in water, bovine serum, and human serum. Moreover, hydrophilic and lipophilic probe release profiles were obtained in HEPES and in human serum. In conclusion, both systems evidenced the entrapment of AgNPs: NioTw20-AgNPs and NioSp20-AgNPs. The two systems are stable in water, bovine serum, and human serum, and maintain the ability to entrap also hydrophilic or lipophilic model molecules. This work demonstrates that the niosomes' features are not altered by AgNPs loading, and confirms that these niosomal formulations are good candidates for the delivery of AgNPs together with other drugs, opening new promising ways for their biotechnological applications.

Supplementary Materials: The following are available online at http://www.mdpi.com/2079-4991/9/8/1177/s1, Figure S1: (**a**) Synthetic scheme for AgNPs-3MPS; (**b**) Uv-vis spectrum of AgNPs-3MPS in water with SPR at $\lambda = 425$ nm; (**c**) $<2R_H> = 8 \pm 3$ nm of AgNPs-3MPS in water; (**d**) ζ-potential = -35 ± 2 mV of AgNPs-3MPS in water; Figure S2: Calibration curve of AgNPs; SI Details about Small angle X-ray Scattering (SAXS).

Author Contributions: Conceptualization, C.M., M.C., I.V. and I.F.; Chemical synthesis and characterizations of AgNPs: I.V. and I.F.; Preparation of niosomes and silver doped niosomes C.M., F.R., P.N.H., M.C.; SAXS investigations: E.d.F., J.M.; AFM investigations: D.P., M.R., L.A.; All authors reviewed and approved the entire manuscript.

Funding: This research received no external funding.

Acknowledgments: The Grant of Excellence Departments, MIUR (ARTICOLO 1, COMMI 314–337 LEGGE 232/2016), is gratefully acknowledged by Iole Venditti. This work has been supported by Italian Ministry of Education, Universities and Research—Dipartimenti di Eccellenza—L. 232/2016 (Department of Drug Chemistry and Technologies, Carafa and Marianecci).

Conflicts of Interest: The authors declare no conflict of interest.

References

1. Clebak, K.T.; Malone, M.A. Skin Infections. *Prim. Care* **2018**, *45*, 433–454. [CrossRef]
2. Manaresi, E.; Gallinella, G. Advances in the Development of Antiviral Strategies against Parvovirus B19. *Viruses* **2019**, *11*, 659. [CrossRef]
3. Hutnick, M.A.; Pokorski, J.K. Polymeric Interventions for Microbial Infections: A Review. *Mol. Pharm.* **2018**, *158*, 2910–2921. [CrossRef]
4. Friedman, D.Z.P.; Schwartz, I.S. Emerging Fungal Infections: New Patients, New Patterns, and New Pathogens. *J. Fungi* **2019**, *5*, 67. [CrossRef]
5. Bunschoten, A.; Welling, M.M.; Termaat, M.F.; Sathekge, M.; van Leeuwen, F.W.B. Development and Prospects of Dedicated Tracers for the Molecular Imaging of Bacterial Infections. *Bioconjug. Chem.* **2013**, *24*, 1971–1989. [CrossRef]
6. Nordøy, I.; Hesstvedt, L.; Torp Andersen, C.; Mylvaganam, H.; Kols, N.I.; Falch, B.M.; Tofteland, S.; Müller, F.; Denning, D.W. An Estimate of the Burden of Fungal Disease in Norway. *J. Fungi* **2018**, *4*, 29. [CrossRef]
7. Wagner, S.; Sommer, R.; Hinsberger, S.; Lu, C.; Hartmann, R.W.; Empting, M.; Titz, A. Novel Strategies for the Treatment of Pseudomonas aeruginosa Infections. *J. Med. Chem.* **2016**, *59*, 5929–5969. [CrossRef]
8. Frieria, M.; Kumar, K.; Boutin, A. Antibiotic resistance. *J. Infect. Public Health* **2017**, *10*, 369–378. [CrossRef]
9. Singh, A.K.; Das, S.; Singh, S.; Gajamer, V.R.; Pradhan, N.; Lepcha, Y.D.; Tiwari, H.K. Prevalence of antibiotic resistance in commensal Escherichia coli among the children in rural hill communities of Northeast India. *PLoS ONE* **2018**, *13*, e0199179. [CrossRef]

10. Mu, H.; Tang, J.; Liu, Q.; Sun, C.; Wang, T.; Duan, J. Potent antibacterial nanoparticles against biofilm and intracellular bacteria. *Sci. Rep.* **2016**, *6*, 18877. [CrossRef]
11. Venditti, I. Engineered gold-based nanomaterials: Morphologies and functionalities in biomedical applications. A mini review. *Bioengineering* **2019**, *6*, 53. [CrossRef]
12. Porcaro, F.; Battocchio, C.; Antoccia, A.; Fratoddi, I.; Venditti, I.; Fracassi, A.; Luisetto, I.; Russo, M.; Polzonetti, G. Synthesis of functionalized gold nanoparticles capped with 3-mercapto-1-propansulfonate and 1-thioglucose mixed thiols and "in vitro" bioresponse. *Colloids Surf. B Biointerfaces* **2016**, *142*, 408–416. [CrossRef]
13. Porcaro, F.; Carlini, L.; Ugolini, A.; Visaggio, D.; Visca, P.; Fratoddi, I.; Venditti, I.; Meneghini, C.; Simonelli, L.; Marini, C. Synthesis and structural characterization of silver nanoparticles stabilized with 3-mercapto-1-propansulfonate and 1-thioglucose mixed thiols for antibacterial applications. *Materials* **2016**, *9*, 1028. [CrossRef]
14. Hong, W.; Zhao, Y.; Guo, Y.; Huang, C.; Qiu, P.; Zhu, J.; Chu, X.; Shi, H.; Liu, M. PEGylated Self-Assembled Nano-Bacitracin A: Probing the Antibacterial Mechanism and Real-Time Tracing of Target Delivery in Vivo. *ACS Appl. Mater. Interfaces* **2018**, *10*, 10688–10705. [CrossRef]
15. Fratoddi, I.; Benassi, L.; Botti, E.; Vaschieri, C.; Venditti, I.; Bessar, H.; Samir, M.A.; Azzoni, P.; Magnoni, C.; Costanzo, A. Effects of topical methotrexate loaded gold nanoparticle in cutaneous inflammatory mouse model. *Nanomedicine* **2019**, *17*, 276–286. [CrossRef]
16. Venditti, I.; Testa, G.; Sciubba, F.; Carlini, L.; Porcaro, F.; Meneghini, C.; Mobilio, S.; Battocchio, C.; Fratoddi, I. Hydrophilic metal nanoparticles functionalized by 2-Diethylaminoethanethiol: A close look at the metal–ligand interaction and interface chemical structure. *J. Phys. Chem. C* **2017**, *121*, 8002–8013. [CrossRef]
17. Prosposito, P.; Mochi, F.; Ciotta, E.; Casalboni, M.; De Matteis, F.; Venditti, I.; Fontana, L.; Testa, G.; Fratoddi, I. Hydrophilic silver nanoparticles with tunable optical properties: Application for the detection of heavy metals in water. *Beilstein J. Nanotechnol.* **2016**, *7*, 1654–1661. [CrossRef]
18. Mochi, F.; Burratti, L.; Fratoddi, I.; Venditti, I.; Battocchio, C.; Carlini, L.; Iucci, G.; Casalboni, M.; De Matteis, F.; Casciardi, S.; et al. Interaction of colloidal silver nanoparticles with Co^{2+} and Ni^{2+} in water for sensing application. *Nanomaterials* **2018**, *8*, 488. [CrossRef]
19. Corsi, P.; Venditti, I.; Battocchio, C.; Meneghini, C.; Bruni, F.; Prosposito, P.; Mochi, F.; Capone, B. Designing an Optimal Ion Adsorber at the Nanoscale: The Unusual Nucleation of AgNP/Co^{2+}–Ni^{2+} Binary Mixtures. *J. Phys. Chem. C* **2019**, *123*, 3855–3860. [CrossRef]
20. Franci, G.; Falanga, A.; Galdiero, S.; Palomba, L.; Rai, M.; Morelli, G.; Galdiero, M. Silver Nanoparticles as Potential Antibacterial Agents. *Molecules* **2015**, *20*, 8856–8874. [CrossRef]
21. Li, Y.; Yang, C.; Yin, X.; Sun, Y.; Weng, J.; Zhou, J.; Feng, B. Inflammatory responses to micro/nano-structured titanium surfaces with silver nanoparticles in vitro. *J. Mater. Chem. B* **2019**, *7*, 3546–3559. [CrossRef]
22. Burdușel, A.C.; Gherasim, O.; Grumezescu, A.M.; Mogoantă, L.; Ficai, A.; Andronescu, E. Biomedical Applications of Silver Nanoparticles: An Up-to-Date Overview. *Nanomaterials* **2018**, *8*, 681. [CrossRef]
23. Aazam, E.S.; Zaheer, Z. Growth of Ag-nanoparticles in an aqueous solution and their antimicrobial activities against Gram positive, Gram negative bacterial strains and Candida fungus. *Bioprocess Biosyst. Eng.* **2016**, *39*, 575–584. [CrossRef]
24. Fayaz, A.M.; Balaji, K.; Girilal, M.; Techc, R.Y.M.; Kalaichelvan, P.T.; Venketesan, R. Biogenic synthesis of silver nanoparticles and their synergistic effect with antibiotics: A study against gram-positive and gram-negative bacteria. *Nanomedicine* **2010**, *6*, 103–109. [CrossRef]
25. Tang, S.; Zheng, J. Antibacterial Activity of Silver Nanoparticles: Structural Effects. *Adv. Healthc. Mater.* **2018**, *7*, e1701503. [CrossRef]
26. Jinu, U.; Jayalakshmi, N.; Sujima Anbu, A.; Mahendran, D.; Sahi, S.; Venkatachalam, P. Biofabrication of Cubic Phase Silver Nanoparticles Loaded with Phytochemicals from Solanum nigrum Leaf Extracts for Potential Antibacterial, Antibiofilm and Antioxidant Activities Against MDR Human Pathogens. *J. Clust. Sci.* **2017**, *28*, 489–505. [CrossRef]
27. Gopinath, P.M.; Narchonai, G.; Dhanasekaran, D.; Ranjani, A.; Thajuddin, N. Mycosynthesis, characterization and antibacterial properties of AgNPs against multidrug resistant (MDR) bacterial pathogens of female infertility cases. *Asian J. Pharm. Sci.* **2015**, *10*, 138–145. [CrossRef]

28. Sanyasi, S.; Majhi, R.K.; Kumar, S.; Mishra, M.; Ghosh, A.; Suar, M.; Satyam, P.V.; Mohapatra, H.; Goswami, C.; Goswami, L. Polysaccharide-capped silver Nanoparticles inhibit biofilm formation and eliminate multi-drug-resistant bacteria by disrupting bacterial cytoskeleton with reduced cytotoxicity towards mammalian cells. *Sci. Rep.* **2016**, *6*, 24929. [CrossRef]
29. Shahverdi, A.R.; Fakhimi, A.; Shahverdi, H.R.; Minaian, S. Synthesis and effect of silver nanoparticles on the antibacterial activity of different antibiotics against Staphylococcus aureus and Escherichia coli. *Nanomedicine* **2007**, *3*, 168–171. [CrossRef]
30. Maneerung, T.; Tokura, S.; Rujiravanit, R. Impregnation of silver nanoparticles into bacterial cellulose for antimicrobial wound dressing. *Carbohydr. Polym.* **2008**, *72*, 43–51. [CrossRef]
31. Corrêa, J.M.; Mori, M.; Sanches, H.L.; Cruz, A.D.D.; Poiate, E.; Poiate, I.A.V.P. Silver nanoparticles in dental biomaterials. *Int. J. Biomater.* **2015**, *2015*, 485275. [CrossRef]
32. Knetsch, M.L.; Koole, L.H. New strategies in the development of antimicrobial coatings: The example of increasing usage of silver and silver nanoparticles. *Polymers* **2011**, *3*, 340–366. [CrossRef]
33. Samuel, U.; Guggenbichler, J. Prevention of catheter-related infections: The potential of a new nano-silver impregnated catheter. *Int. J. Antimicrob. Agents* **2004**, *23*, 75–78. [CrossRef]
34. Zhang, X.-F.; Liu, Z.-G.; Shen, W.; Gurunathan, S. Silver nanoparticles: Synthesis, characterization, properties, applications, and therapeutic approaches. *Int. J. Mol. Sci.* **2016**, *17*, 1534. [CrossRef]
35. Ran, L.; Zou, Y.; Cheng, J.; Lu, F. Silver nanoparticles in situ synthesized by polysaccharides from Sanghuangporus sanghuang and composites with chitosan to prepare scaffolds for the regeneration of infected full-thickness skin defects. *Int. J. Biol. Macromol.* **2019**, *125*, 392–403. [CrossRef]
36. Panáček, A.; Kvítek, L.; Smékalová, M.; Večeřová, R.; Kolář, M.; Röderová, M.; Dyčka, F.; Šebela, M.; Prucek, R.; Tomanec, O.; et al. Bacterial resistance to silver nanoparticles and how to overcome it. *Nat. Nanotechnol.* **2018**, *13*, 65–71. [CrossRef]
37. Elbehiry, A.; Al-Dubaib, M.; Marzouk, E.; Moussa, I. Antibacterial effects and resistance induction of silver and gold nanoparticles against Staphylococcus aureus-induced mastitis and the potential toxicity in rats. *MicrobiologyOpen* **2019**, *8*, e698. [CrossRef]
38. Das, A.; Debasis, S.; Jitendra, B.; Mahesh, S. Confinement induced formation of silver nanoparticles in self-assembled micro-granules. *Colloids Surf. A Physicochem. Eng. Asp.* **2019**, *577*, 185–193. [CrossRef]
39. Huang, J.; Zhou, Y.-f.; Xu, J.; Liang, P.; Liu, Z.-g.; Wang, J.; Zhang, D.; Dong, Q.-m.; Shen, W.-m.; Zhuang, S.-l. Unveiling the growth mechanism of SiO2/Ag hybrid nanospheres and using for Surface Enhanced Raman Scattering detection. *Appl. Surf. Sci.* **2019**, *463*, 115–120. [CrossRef]
40. Kumar, K.A.; John, J.; Sooraj, T.; Raj, S.A.; Unnikrishnan, N.; Selvaraj, N.B. Surface plasmon response of silver nanoparticles doped silica synthesised via sol-gel route. *Appl. Surf. Sci.* **2019**, *472*, 40–45. [CrossRef]
41. Lin, L.; Zhang, H.; Cui, H.; Xu, M.; Cao, S.; Zheng, G.; Dong, M. Preparation and antibacterial activities of hollow silica–Ag spheres. *Colloids Surf. B Biointerfaces* **2013**, *101*, 97–100. [CrossRef]
42. Nishanthi, S.; Yadav, K.K.; Baruah, A.; Vaghasiya, K.; Verma, R.K.; Ganguli, A.K.; Jha, M. Nanostructured silver decorated hollow silica and their application in the treatment of microbial contaminated water at room temperature. *New J. Chem.* **2019**, *43*, 8993–9001. [CrossRef]
43. Gao, A.; Chen, H.; Hou, A.; Xie, K. Efficient antimicrobial silk composites using synergistic effects of violacein and silver nanoparticles. *Mater. Sci. Eng. C* **2019**, *103*, 109821. [CrossRef]
44. Wang, Y.; Wang, Y.; Su, L.; Luan, Y.; Du, X.; Zhang, X. Effect of surface topology morphologies of silica nanocarriers on the loading of Ag nanoparticles and antibacterial performance. *J. Alloys Compd.* **2019**, *783*, 136–144. [CrossRef]
45. Marianecci, C.; Di Marzio, L.; Rinaldi, F.; Celia, C.; Paolino, D.; Alhaique, F.; Esposito, S.; Carafa, M. Niosomes from 80s to present: The state of the art. *Adv. Colloid Interface Sci.* **2014**, *205*, 187–206. [CrossRef]
46. Venditti, I. Morphologies and functionalities of polymeric nanocarriers as chemical tools for drug delivery: A review. *J. King Saud Univ. Sci.* **2019**, *31*, 398–411. [CrossRef]
47. Ge, X.; Wei, M.; He, S.; Yuan, W.-E. Advances of Non-Ionic Surfactant Vesicles (Niosomes) and Their Application in Drug Delivery. *Pharmaceutics* **2019**, *11*, 55. [CrossRef]
48. Yusuf, A.; Brophy, A.; Gorey, B.; Casey, A. Liposomal encapsulation of silver nanoparticles enhances cytotoxicity and causes induction of reactive oxygen species-independent apoptosis. *J. Appl. Toxicol.* **2018**, *38*, 616–627. [CrossRef]

49. Rinaldi, F.; Seguella, L.; Gigli, S.; Hanieh, P.; Del Favero, E.; Cantù, L.; Pesce, M.; Sarnelli, G.; Marianecci, C.; Esposito, G. In Pentasomes: An innovative nose-to-brain pentamidine delivery blunts MPTP parkinsonism in mice. *J. Control. Release* **2019**, *294*, 17–26. [CrossRef]
50. Rinaldi, F.; Hanieh, P.N.; Chan, L.K.N.; Angeloni, L.; Passeri, D.; Rossi, M.; Wang, J.T.-W.; Imbriano, A.; Carafa, M.; Marianecci, C. Chitosan Glutamate-Coated Niosomes: A Proposal for Nose-to-Brain Delivery. *Pharmaceutics* **2018**, *10*, 38. [CrossRef]
51. De Angelis, R.; Venditti, I.; Fratoddi, I.; De Matteis, F.; Prosposito, P.; Cacciotti, I.; D'Amico, L.; Nanni, F.; Yadav, A.; Casalboni, M.; et al. From nanospheres to microribbons: Self-assembled Eosin Y doped PMMA nanoparticles as photonic crystals. *J. Colloid Interf. Sci.* **2014**, *414*, 24–32. [CrossRef]
52. Sennato, S.; Bordi, F.; Cametti, C.; Marianecci, C.; Carafa, M.; Cametti, M. Hybrid niosome complexation in the presence of oppositely charged polyions. *J. Phys. Chem. B* **2008**, *112*, 3720–3727. [CrossRef]
53. Rao, H.S.P.; Desai, A.; Sarkar, I.; Mohapatra, M.; Mishra, A.K. Photophysical behavior of a new cholesterol attached coumarin derivative and fluorescence spectroscopic studies on its interaction with bile salt systems and lipid bilayer membranes. *Phys. Chem. Chem. Phys.* **2014**, *16*, 1247–1256. [CrossRef]
54. Lakowicz, J.R. *Principles of Fluorescence Spectroscopy*; Springer Science & Business Media: Berlin/Heidelberg, Germany, 2013.
55. Lentz, B.R. Membrane "fluidity" as detected by diphenylhexatriene probes. *Chem. Phys. Lipids* **1989**, *50*, 171–190. [CrossRef]
56. Zachariasse, K.A. Intramolecular excimer formation with diarylalkanes as a microfluidity probe for sodium dodecyl sulphate micelles. *Chem. Phys. Lett.* **1978**, *57*, 429–432. [CrossRef]
57. Ingallina, C.; Rinaldi, F.; Bogni, A.; Ponti, J.; Passeri, D.; Reggente, M.; Rossi, M.; Kinsner-Ovaskainen, A.; Mehn, D.; Rossi, F. Niosomal approach to brain delivery: Development, characterization and in vitro toxicological studies. *Int. J. Pharm.* **2016**, *511*, 969–982. [CrossRef]
58. Rinaldi, F.; Hanieh, P.N.; Del Favero, E.; Rondelli, V.; Brocca, P.; Pereira, M.C.; Andreev, O.A.; Reshetnyak, Y.K.; Marianecci, C.; Carafa, M. Decoration of nanovesicles with pH (low) insertion peptide (pHLIP) for targeted delivery. *Nanoscale Res. Lett.* **2018**, *13*, 391. [CrossRef]
59. Carafa, M.; Marianecci, C.; Rinaldi, F.; Santucci, E.; Tampucci, S.; Monti, D. Span®and Tween®neutral and pH-sensitive vesicles: Characterization and in vitro skin permeation. *J. Liposome Res.* **2009**, *19*, 332–340. [CrossRef]
60. Marianecci, C.; Di Marzio, L.; Del Favero, E.; Cantù, L.; Brocca, P.; Rondelli, V.; Rinaldi, F.; Dini, L.; Serra, A.; Decuzzi, P. Niosomes as drug nanovectors: Multiscale pH-dependent structural response. *Langmuir* **2016**, *32*, 1241–1249. [CrossRef]
61. Palchetti, S.; Colapicchioni, V.; Digiacomo, L.; Caracciolo, G.; Pozzi, D.; Capriotti, A.L.; La Barbera, G.; Laganà, A. The protein corona of circulating PEGylated liposomes. *Biochim. Biophys. Acta* **2016**, *1858*, 189–196. [CrossRef]

MDPI
St. Alban-Anlage 66
4052 Basel
Switzerland
Tel. +41 61 683 77 34
Fax +41 61 302 89 18
www.mdpi.com

Nanomaterials Editorial Office
E-mail: nanomaterials@mdpi.com
www.mdpi.com/journal/nanomaterials

www.ingramcontent.com/pod-product-compliance
Lightning Source LLC
Chambersburg PA
CBHW062103220926
51850CB00010B/2867
* 9 7 8 3 0 3 9 2 8 8 3 3 5 *